International Workshop on Nonlinear Maps and Applications (NOMA15)

Journal of Physics: Conference Series Volume 692

Dublin, Ireland
15 - 16 June 2015

ISBN: 978-1-5108-2185-9
ISSN: 1742-6588

Printed from e-media with permission by:

Curran Associates, Inc.
57 Morehouse Lane
Red Hook, NY 12571

Some format issues inherent in the e-media version may also appear in this print version.

Printed by Curran Associates, Inc. (2016)

For permission requests, please contact the Institute of Physics
at the address below.

Institute of Physics
Dirac House, Temple Back
Bristol BS1 6BE UK

Phone: 44 1 17 929 7481
Fax: 44 1 17 920 0979

techtracking@iop.org

Additional copies of this publication are available from:

Curran Associates, Inc.
57 Morehouse Lane
Red Hook, NY 12571 USA
Phone: 845-758-0400
Fax: 845-758-2633
Email: curran@proceedings.com
Web: www.proceedings.com

Table of contents

Volume 692

February 2016

NOMA'15 International Workshop on Nonlinear Maps and Applications / 15–16 June 2015, Dublin, Ireland

Accepted papers received: 10 February 2016 Published online: 14 March 2016

Abstract

The International Workshop on Nonlinear Maps and their Applications (NOMA) is a series of international conferences. NOMA editions were held in Toulouse (Noma'07), Urbino (Noma'09), Évora (Noma'11) and Zaragoza (Noma'13). The fifth edition of NOMA was organised and hosted by the School of Electrical and Electronic Engineering of University College Dublin. This workshop brings together researchers from theoretical and application areas (mathematics, physics, engineering and economics) who study nonlinear discrete systems.

Nonlinear iterative processes play an important role in physical, biological and social phenomena. Nonlinear mappings can directly model various systems or can be obtained using numerical methods permitting the solution of differential nonlinear equations. In both cases, the understanding of specific behaviours and bifurcations of these type of systems is of the greatest interest. This workshop is open to theoretical studies as well as applicative ones in the fields of physics, electronics, biology, computational methods, engineering, telecommunications and others.

The scientific programme of NOMA'15 included 28 invited and regular lectures with 12 selected talks published in this special issue. Prof Vassili Gelfreich (the University of Warwick), Prof Daniele Fournier-Prunaret (INSA Toulouse), Prof Ricardo Lopez-Ruiz (the University of Zaragoza), Prof Sergio Callegari (the University of Bologna), Prof Yoshifumi Nishio (Tokushima University) and Dr Elena Blokhina (University College Dublin) have served as the editors of NOMA'2015 and selected the papers. On behalf of the scientific committee of NOMA, we would like to thank the editors and Eoghan O'Riordan and Panagiotis Giounanlis for their help in preparing this special issue. We are very grateful for the support of University College Dublin and to the School of Electrical and Electronic Engineering.

Peer review statement

All papers published in this volume of *Journal of Physics: Conference Series* have been peer reviewed through processes administered by the proceedings Editors. Reviews were conducted by expert referees to the professional and scientific standards expected of a proceedings journal published by IOP Publishing.

Are numerical studies of long term dynamics conclusive: the case of the Hénon map

Zbigniew Galias

AGH University of Science and Technology, Department of Electrical Engineering, al. Mickiewicza 30, 30–059 Kraków, Poland

E-mail: galias@agh.edu.pl

Abstract. Numerical studies of the Hénon map show that there is an abundance of periodic windows close to classical parameter values. In this work properties of these periodic windows are studied. It is shown that in order to detect periodic windows in simulations one has to compute trajectories in a sufficient precision, which in many cases is much higher than the standard double precision. Moreover, it is shown that even if computations are carried out in a sufficient precision, one may need an extremely large number of iterations to observe convergence to a sink starting from random initial conditions, and hence the underlying steady state behavior may be practically undetectable using the standard trajectory monitoring based approach.

1. Introduction

Numerical simulations play an important role in studying dynamics of nonlinear systems. Simulations are used to compute long trajectories to study steady-state behaviour, to provide data for bifurcation diagrams, to compute Lyapunov exponents to classify attractors, etc. However, due to inherent properties of digital computers, results found by numerical simulations are almost never exact [1, 2]. Nevertheless, computer generated solutions are often accepted as true solutions.

In this work, we will show examples that simulation studies of nonlinear systems as simple as the Hénon map cannot, in general, provide conclusive answer to the question what is the true nature of the attractor existing for fixed parameter values. It follows that what we observe in many examples reported in the literature, and what is claimed to be a chaotic trajectory, might in fact be a transient to a periodic steady-state or an artifact caused by rounding errors.

2. Detection of periodic windows for the Hénon map

The *Hénon map* [3] is a two-parameter, invertible map of a plane into itself

$$h(x, y) = (1 + y - ax^2, bx), \tag{1}$$

displaying a wide array of dynamical behaviors as its parameters are varied. In [3], the map h is numerically studied with parameter values $(a, b) = (1.4, 0.3)$, referred to in the following as the *classical parameter values*, and it is claimed that "depending on the initial point, the sequence of points obtained by iteration of the mapping either diverges to infinity or tends to a strange attractor". For the classical parameter values the so-called *Hénon attractor* is observed in simulations.

Fig. 1(a) shows a bifurcation diagram for the Hénon map. The value of b is fixed ($b = 0.3$) and 10000 values of a in the range $a \in [0, 1.42]$ are selected. For each parameter value 10000 iterates are skipped and the x variable for the next 50000 iterates is plotted. Period-doubling route to chaos and low-period windows in the chaotic range are clearly visible.

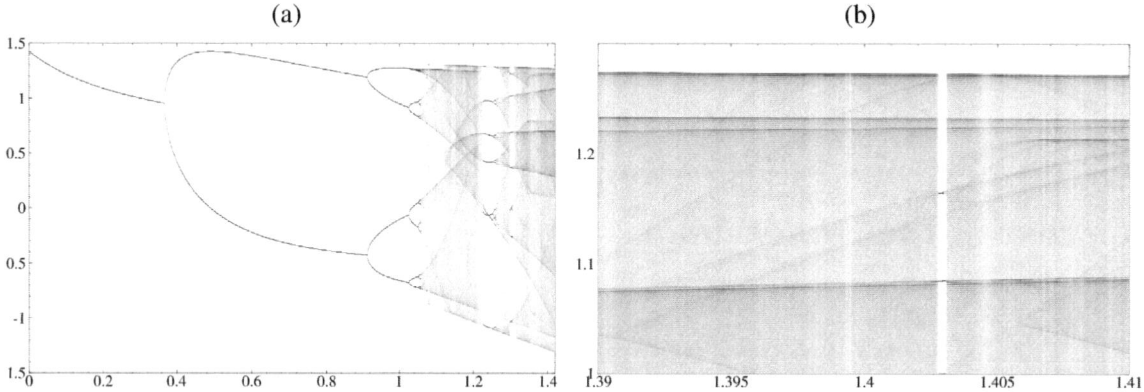

Figure 1. Bifurcation diagram for the Hénon map, $x \in [-1.5, 1.5]$ versus a; $b = 0.3$, (a) 10000 parameter values $a \in [0, 1.42]$, 10000 iterates skipped, 50000 iterates plotted, (b) 10000 parameter values $a \in [1.39, 1.41]$, 100000 iterates skipped, 50000 iterates plotted, $x \in [1.0, 1.3]$.

A bifurcation diagram for a smaller interval $a \in [1.39, 1.41]$ is shown in Fig. 1(b). The number of skipped iteration has been increased to 100000. One can see a single wide periodic window and a few narrower ones (not clearly visible). Important questions are how many periodic windows can be detected using selected grid of parameter values, and whether the number of skipped iterations in a given periodic window is sufficient to converge to the corresponding sink. In the remaining part of the paper we will discuss these problems in detail.

The most natural way to find periodic windows is to using the monitoring trajectory based approach. In this method, a number of iterates are computed in the hope that a trajectory reaches a steady-state. Next, we take the current iterate as the new initial point and check if the trajectory periodically returns very close to this point. This method has been used in [4] to find periodic windows in a small neighborhood of the classical parameter values. In an example search test 10^8 points in the parameter space filling uniformly the interval $[1.3999, 1.4001] \times 0.3$ have been considered. For each point 1000 random initial conditions have been selected and for each initial condition 10^7 iterations have been computed to detect stable periodic orbits. Note that the total number of iterations needed to carry out such a test is 10^{17}. In spite of very long computation time, only 461 periodic windows intersecting the region $Q_0 = [1.3999, 1.4001] \times [0.2999, 0.3001]$ have been found. This failure is caused by a number of factors. First, detection of periodic windows may require arithmetic precision higher than standard double-precision. Second, periodic windows may be too narrow to be detected with a chosen sampling of the parameter space. Third, the number of iterations to observe the convergence may be very large.

A more systematic approach to detect periodic windows has been proposed in [5]. It is based on finding a carefully selected set of periodic orbits existing for fixed parameter values, and then using continuation in the parameter space to find parameter values sustaining a sink. In this method, two close points in the parameter space are selected and for each of them the existence of periodic orbits corresponding to specific symbol sequences is verified using the Biham-Wenzel method [6]. Symbol sequences admissible for only one of the two considered points correspond to periodic windows intersecting the interval defined by these points. Positions of periodic windows are found using the continuation technique. We start at the endpoint where the orbit exist, modify parameter values moving towards the second endpoint where the orbit is not admissible. When moving along the interval in the parameter space, the position of the orbit is updated, eigenvalues of the corresponding Jacobian matrix are calculated, and stability conditions are verified. When a point in parameter space with a stable periodic behavior is found, the calculations are stopped with a candidate of a point belonging to the

Journal of Physics: Conference Series **692** (2016) 012001 doi:10.1088/1742-6596/692/1/012001

periodic window. The existence of a stable periodic orbit is verified using the interval Newton method. For more details see [5]. Selection of symbol sequences to be considered for a given sequence length is based on results obtained for the lower-period windows. This step is necessary since considering all symbol sequences for larger periods is not feasible.

Using this method, many (tens of thousands) low-period windows very close to (1.4, 0.3) have been found. These include several periodic windows within the distance 10^{-20} from the classical parameter values. The closest one is the period-115 window at the distance less than 7×10^{-22} from the point (1.4, 0.3).

3. Convergence properties

In this section, we study properties of periodic windows intersecting a small neighborhood of the classical parameter values $(a, b) = (1.4, 0.3)$. Computation are carried out in the double-precision arithmetic or in multiple-precision arithmetic using the MPFR library [7]. Interface to multiple-precision interval arithmetic calculations is provided by the CAPD library [8].

Fig. 2 shows a trajectory of the Hénon map with parameters $a = 1.400000009849371$ and $b = 0.300000019143266$ for the initial point $(x_0, y_0) = (0, 0)$ computed in the standard double precision arithmetic. 4×10^{11} iterations are skipped in the hope that the steady-state has been reached, and the next 10000 points are plotted using red dots. One can see that the plot is visually indistinguishable from the well-known Hénon attractor. However, if we wait longer the trajectory converges to the period-31 stable periodic orbit plotted as blue circles in Fig. 2.

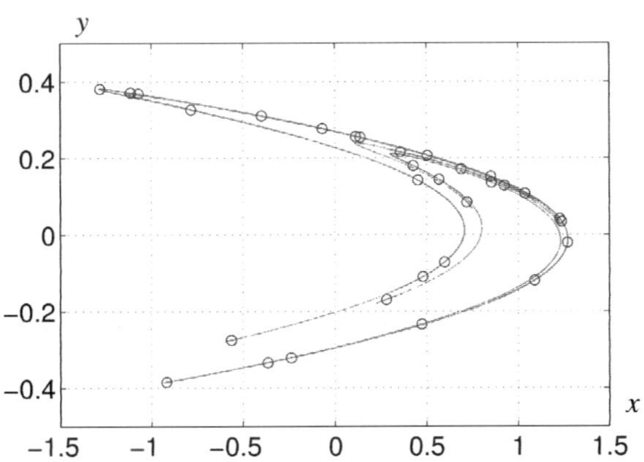

Figure 2. Trajectory of the Hénon map computed in double-precision, $a = 1.400000009849371$, $b = 0.300000019143266$, $(x_0, y_0) = (0, 0)$, 10000 points after skipping 4×10^{11} iterations are plotted using red dots, 10000 points after skipping 5×10^{11} iterations are plotted using blue circles.

The parameter values considered belong to one of the periodic windows detected using the symbolic dynamics based approach [5]. The existence of the stable periodic orbit has been rigorously confirmed using the interval Newton method [9, 10]. Therefore, we can be sure that the observed periodic behaviour is not a rounding error artifact. We have tested various initial conditions and confirmed in simulations that all trajectories converge to the same period-31 sink, which indicates that this sink is perhaps the only attractor existing for the considered parameter values. This example shows that finite pieces of trajectories, even if they are very long, are inconclusive as to the true nature of the steady-state of the system. It follows that we need to be very cautious drawing any conclusions from study of such finite trajectories. For example from the fact that the largest Lyapunov exponent computed along a transient trajectory is positive it does not follow that the steady state is chaotic. In the example considered, the maximum Lyapunov exponent becomes negative only after the steady-state is reached.

Let us assume that the considered parameter values (a, b) belong to a periodic window, i.e. there exist a stable periodic orbit for these parameter values. One can show that sufficiently close to the classical parameter values the Hénon map has infinitely many periodic orbits. This follows from the existence of nontrivial symbolic dynamics with positive topological entropy [11]. Most of these orbits are unstable.

It follows that even if parameter values belong to a periodic window and the sink is the only attractor for the system, it may take an arbitrarily large number of iterations to observe convergence to this sink. Given the iteration number n and any of the unstable periodic orbits one may choose an initial point sufficiently close to this unstable orbit so that only after more than n iterations a trajectory leaves a small neighborhood of the unstable periodic orbit and then it may converge to the sink. An interesting property of stable periodic orbits is the average convergence time starting from random initial conditions. If the average convergence time is large then we have little chance of finding the sink observing short trajectories. Let $n_{\text{conv}}(\varrho)$ denote the number of iterations which are required to converge to the sink with probability $\varrho \in (0, 1)$ starting from random initial conditions. To approximate $n_{\text{conv}}(\varrho)$, we first compute trajectories for N random initial points and record the corresponding convergence times $n_1, n_2, \ldots, n_N$. Next, we sort the convergence times $n_1 \leq n_2 \leq \cdots \leq n_N$ and find k such that $k \approx \varrho N$. n_k serves as an approximation of $n_{\text{conv}}(\varrho)$. In the following, we will use $\varrho = 0.5$.

Later, it will be shown that it is difficult to get a reliable approximation of n_{conv} for parameter values close to the classical case. To investigate convergence properties we will use the notion of the immediate basin of attraction and its radius [4]. Let $A = (z_0, z_1, \ldots, z_{p-1})$ be a periodic attractor. We say that a point z belongs to the *immediate basin of attraction* $\mathrm{B}_\varepsilon(A)$ of the attractor A if its trajectory converges to the attractor and does not escape further than ε from it, i.e.

$$\mathrm{B}_\varepsilon(A) = \left\{ z \colon \|(h^n(z) - h^n(z_k)\| \leq \varepsilon \; \forall n \geq 0 \text{ and } \lim_{n\to\infty} \|(h^n(z) - h^n(z_k)\| = 0 \text{ for some } 0 \leq k < p \right\}. \tag{2}$$

We will use ε to be 1% of the diameter of the smallest ball enclosing the attractor. The condition $z \in \mathrm{B}_\varepsilon(A)$ can be verified numerically by computing the trajectory of z in arithmetic of sufficient precision and verifying whether conditions in (2) hold. Computing accurately the measure $\mu(\mathrm{B}_\varepsilon(A))$ of the immediate basin of attraction requires very fine sampling of the state space and is computationally expensive. It is easier to find connected components of $\mathrm{B}_\varepsilon(A)$, each containing one of the points z_k. In this way, the search is limited to grid points in the state space which are connected to points already identified as belonging to $\mathrm{B}_\varepsilon(A)$. To further reduce the computation time needed to find a lower bound of $\mu(\mathrm{B}_\varepsilon(A))$ we will use the notion of the *immediate basin radius of the attractor* A at the point z_k which is defined as

$$\mathrm{r}_\varepsilon(z_k) = \sup\{r \colon \|z - z_k\| \leq r \Rightarrow z \in \mathrm{B}_\varepsilon(A)\}. \tag{3}$$

Computation of $\mathrm{r}_\varepsilon(z_k)$ is straightforward. For example, the bisection method can be used to obtain an accurate approximation of the largest r such that the condition $z \in \mathrm{B}_\varepsilon(A)$ is satisfied for test points in a ball of radius r centered at z_k.

The *minimum immediate basin radius of the attractor* A is defined as

$$\mathrm{r}_\varepsilon(A) = \min_{0 \leq k < p} \mathrm{r}_\varepsilon(z_k). \tag{4}$$

If the minimum immediate basin radius of a sink is smaller than the arithmetic precision used then trajectories will most likely escape from the sink even if computations are started at a stable periodic point; in general we will not be able to observe the sink in simulations (compare [1]). Provided that the computations are carried out in a sufficient precision, a trajectory initiated at a point closer to the sink than its immediate basin radius converges to the sink without leaving its small neighborhood; in simulations we will see a stable periodic behavior.

To illustrate the above definitions let us consider the point $(a, b) = (1.4028, 0.3)$ in the parameter space. This point belongs to a period-9 window. Position of one of the stable periodic points existing for $(a, b) = (1.4028, 0.3)$ is shown in Fig. 3 using the star symbol. The border of the connected component containing this point of the immediate basin of attraction of the periodic orbit is plotted in green. The immediate basin radius at this point is plotted as a red circle. Chaotic transient trajectories are plotted as blue dots.

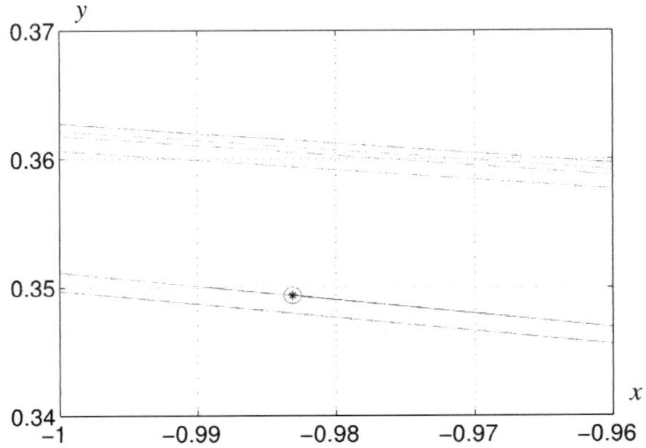

Figure 3. A stable period–9 point existing for $a = 1.4028$, $b = 0.3$ (the star symbol); chaotic transient trajectories (blue dots); border of the connected component of the immediate basin of attraction (green lines).

Convergence properties depend on the measure of the intersection of a chaotic set with the immediate basin. We will show that the measure of the immediate basin and the minimum immediate basin radius can be used to estimate the average time needed for a trajectory to converge to the sink. Let us first consider the case $b = 0$. When $b = 0$, the Hénon map reduces to $h(x) = 1 - ax^2$, which after a linear change of variable is equivalent to the logistic map $f(x) = ax(1 - x)$ (with different a). Let us first study convergence properties for a selected periodic window. For the logistic map each periodic window corresponds to a symbol sequence $s = (s_0, s_1, \ldots, s_{p-1})$, with $s_k \in \{-1, +1\}$, where the number of positive symbols is odd. In [12], it has been shown how to efficiently find the position of a periodic window for a given symbol sequence using the Newton method applied to find bifurcation points. Applying this procedure to the symbol sequence $s = (---+--+--+)$ gives the window's endpoints $(\underline{a}, \overline{a}) \approx (3.96776003545, 3.96776027772)$. Stability of the periodic orbit $(x_0, x_1, \ldots, x_{p-1})$ depends on the value of $\lambda = (f^p)'(x_0) = f'(x_{p-1}) \cdots f'(x_1)f'(x_0)$. If $|\lambda| < 1$ then the orbit is asymptotically stable. Fig. 4(a) shows λ versus a for the periodic window considered. The left and the right endpoints correspond to saddle-node and period-doubling bifurcation points with $\lambda = 1$ and $\lambda = -1$, respectively.

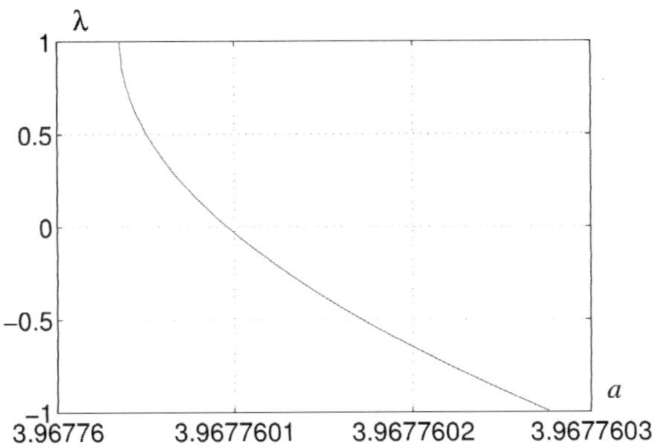

Figure 4. The derivative $\lambda = (f^p)'(x_0)$ across the periodic window of the logistic map corresponding to the symbol sequence $s = (---+--+--+)$.

For the logistic map the state space is one-dimensional and for a given stable periodic orbit $(x_0, x_1, \ldots, x_{p-1})$ it is relatively easy to find p connected components of the immediate basin of attraction of the sink containing points x_k. Let us denote by L the total length of these connected components. The total length L versus the parameter a across the selected periodic window is shown in Fig. 5(a). One can see that when the parameter approaches the saddle-node bifurcation point the total length L goes to zero. This is a consequence of the fact that at the saddle-node point the immediate basin of attraction disappears. Somewhat unexpected is the monotonicity of L across the window. The maximum is observed at the right window's endpoint where the period-doubling bifurcation occurs. One could expect

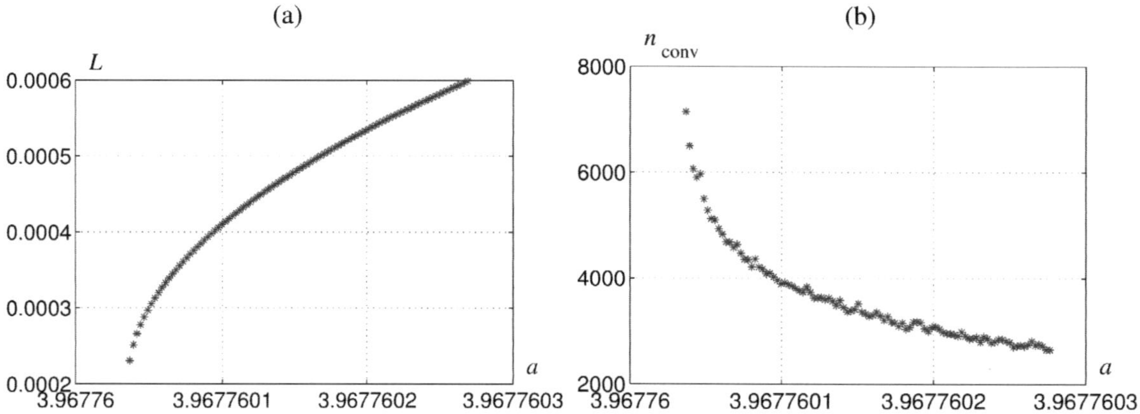

Figure 5. Convergence properties in the periodic window of the logistic map corresponding to the symbol sequence $s = (---+--+--+)$, (a) measure of subset of $B_\varepsilon(A)$ composed of connected components containing points x_k, (b) average number of iterations $n_{\mathrm{conv}}(0.5)$ to converge to the sink.

that when λ is small in magnitude, then the immediate basin of attraction should be large. However, this is not true. The total length L for superstable orbit (with $\lambda = 0$) is smaller than for larger a when λ is nonzero. Average convergence times across the periodic window are shown in Fig. 5(b). These results have been obtained by selecting 100 equidistant parameter values belonging to the periodic window. For each parameter value 10000 trajectories starting from random initial conditions have been computed, convergence time was recorded, and the average convergence time has been found. Generally, the average convergence time decreases when a increases. It escapes to infinity when parameter approaches the saddle-node bifurcation point (left window's endpoint).

Note that there is a strong correlation between L and n_{conv} (compare Fig. 5(a) and (b)). Below, we derive an approximate formula for the relation between L and n_{conv}. We will assume that for parameter values considered there exist a single periodic attractor (for the logistic map it is known that there exists at most one attractor), and that transient trajectories visit the invariant set $\mathcal{A}$ containing all periodic orbits according to the uniform distribution. The true distribution is not flat, but we will see that anyway the resulting formula produces acceptable results. Let us assume that the measure of $\mathcal{A}$ is D. The probability that at least one out of n points selected randomly from the set of measure D belongs to a subset of measure L is $\varrho = 1 - (1 - L/D)^n$. Solving for n with the assumption $L \ll D$ yields

$$n = \frac{\log(1-\varrho)}{\log(1-\frac{L}{D})} \approx \frac{D\log(1-\varrho)^{-1}}{L}. \tag{5}$$

When a is close to 4 the measure of the set $\mathcal{A}$ is close to one. Assuming that $\varrho = 0.5$ yields

$$n \approx 0.693 \cdot L^{-1}. \tag{6}$$

Thus, we can expect that the average number n_{conv} of iterations to converge to a sink whose immediate basin has measure L is inversely proportional to L. To verify this relationship, in Fig. 6 we plot n_{conv} versus L for eleven windows of the logistic map with periods $p \in [9, 26]$. The corresponding window width are in the range $[10^{-15}, 10^{-6}]$. One can see that the plot is composed of short intervals. Each interval corresponds to parameter values belonging to one of the periodic windows considered. For smaller L the plot is more noisy. This is related to the fact that for small window width (close to 10^{-15}) the minimum immediate basin radius of the periodic attractor is also very small. In consequence, the effects caused by finite precision of the double precision arithmetic become visible. Data points in Fig. 6

Journal of Physics: Conference Series **692** (2016) 012001 doi:10.1088/1742-6596/692/1/012001

lie close to a straight line

$$\log n_{\rm conv} = c_1 \log L + c_0, \tag{7}$$

with $c_1 = -0.982$, $c_0 = 0.7642$, which is plotted in Fig. 6 in green. One can see that c_1 is close to -1, which confirms the approximate formula (6). We may use (7) to extrapolate what is the average convergence time $n_{\rm conv}$ for very small L when direct computation of $n_{\rm conv}$ is not feasible. For example when $L = 10^{-30}$ we expect that the average convergence time is around 6×10^{29}. Using one of the fastest computers available nowadays with the performance close to 50 petaflops (5×10^{16} floating point operations per second) with the assumption that three floating point operations are needed to compute one iteration of the logistic map, one would need on average more than one million years of computations to see the convergence.

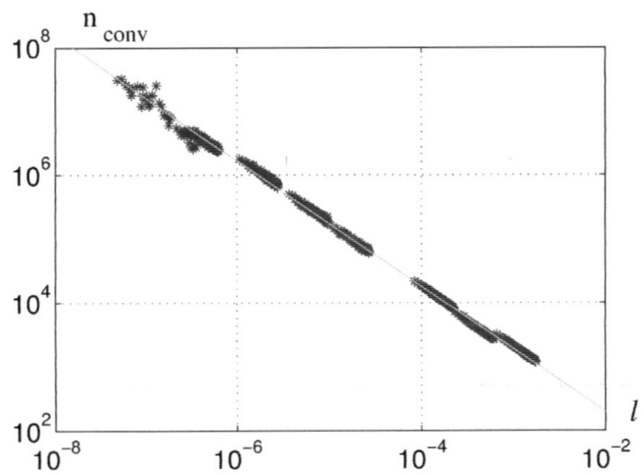

Figure 6. The average number $n_{\rm conv}$ of iterations needed to observe convergence versus the measure L of the immediate basin for several periodic windows of the logistic map.

Let us now consider periodic windows of the Hénon map close to classical parameter values. Fig. 7(a) shows the relationship between the minimum immediate basin radius r_ε and the precision (in bits) needed to see the sink in simulations. This number is computed as the minimum number of bits of multi-precision arithmetic used such that the computer generated trajectory initiated exactly (with the given precision) at the sink position does not escape from a small neighborhood of the sink. One can see that the plot is close to the line: $-\log_2 \mathrm{r}_\varepsilon$ plotted in Fig. 7(a) as a red solid line. This phenomenon can be easily explained in the following way. If the arithmetic precision does not allow us to represent points belonging to a neighborhood of a periodic point with the radius r_ε then most likely a trajectory of a periodic point represented in this precision will escape from this little neighborhood.

To study convergence properties, in Fig. 7(b) we plot the relationship between the minimum immediate basin radius r_ε and average number $n_{\rm conv}$ of iterations to converge to a sink with probability $\varrho = 0.5$. This data was based on computing trajectories starting at 10000 randomly chosen initial conditions for several periodic windows intersecting the interval $(a, b) \in [1.39, 1.41] \times \{0.3\}$. For small r_ε the number of initial points was decreased due to very long computation times. Fitting data points shown in Fig. 7(b) to the model

$$\log n_{\rm conv} = c_1 \log \mathrm{r}_\varepsilon + c_0. \tag{8}$$

gives $c_1 = -0.7589$, $c_0 = 0.0566$. This model is plotted in Fig. 7(b) as a green line. Note that c_1 is significantly smaller than 1 in absolute value in opposite to the logistic map case (compare (7)). This is related to the fact that the chaotic set for the Hénon map is not one-dimensional, but has a complex fractal structure. The results obtained indicate that we may use formula (8) to extrapolate the average return time for periodic orbits with very small minimum immediate basin radius.

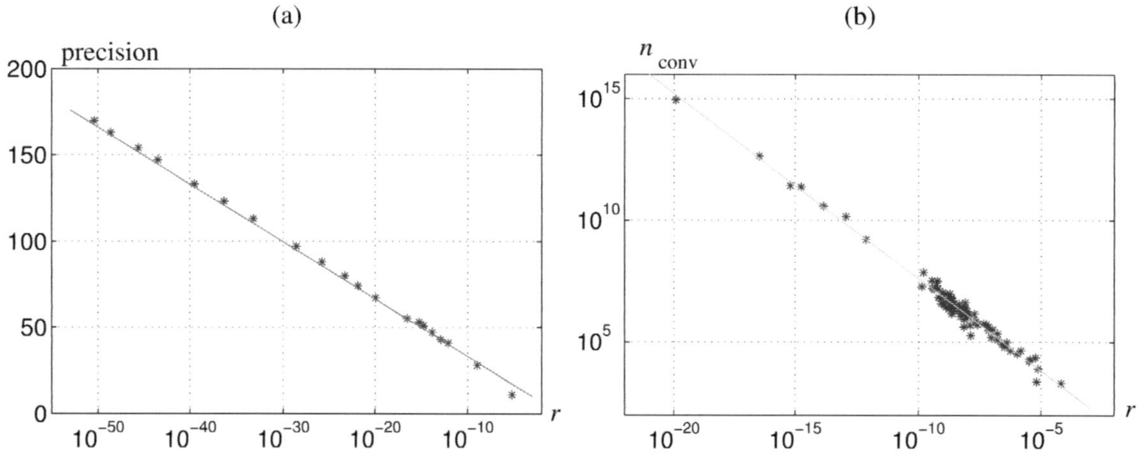

Figure 7. Properties of periodic windows of the Hénon map close to classical parameter values; (a) precision needed to observe the sink versus the minimum immediate basin radius r_ε, (b) The average number n_{conv} of iterations needed to observe convergence versus r_ε.

Let us now apply the results presented above to an example periodic window. Let us consider the closest periodic window reported in [5] with the distance from the point $(1.4, 0.3)$ smaller than 7×10^{-22} and the period equal to 115. Computer arithmetic precision necessary to detect this sink is 170 bits. We verified that if the computations are carried out in lower precision then the trajectory escapes from the sink even if it is started exactly (with the precision used) at the sink position. The minimum immediate basin radius for this sink is $r_\varepsilon \approx 3.72 \times 10^{-51}$. Using formula (8) we obtain $n_{\text{conv}} \approx 2 \times 10^{38}$. We expect that so many iterations are needed to observe with probability $\varrho = 0.5$ the convergence to the corresponding periodic orbit starting from random initial conditions. Assuming that 10^{15} iterations can be computed in one second (which is far beyond capabilities of even fastest supercomputers) one would need more than 10^{16} years to reach $n_{\text{conv}} = 2 \times 10^{38}$. Since it is impractical to compute so many iterations, we conclude that this and similar periodic windows are practically undetectable using simulation based approach even if computations are carried out with the required precision.

4. Conclusions

We have shown that in order to detect in simulation periodic windows lying very close to classical parameter values of the Hénon map one has to use arithmetic precision with a sufficient number of bits, which is usually much larger than the standard 53 bits precision. It follows that the true nature of the corresponding attractors cannot be verified using standard double-precision computations. We have estimated the average number of iterations needed to converge to corresponding sinks starting from random initial conditions. We have shown that this number can be so large (above 10^{30}) even for sinks of moderate length (with period 100 or so) that these sinks are practically undetectable in simulations even if computations are carried out with the required precision. Although in our study we restricted ourselves to the case of a simple two-dimensional map we believe that the results carry over to many other discrete and continuous dynamical systems.

We conclude with the statement that most of the simulation based results concerning chaotic behaviour of nonlinear systems published in the literature are irrelevant to the question what is the true character of the underlying attractor — no conclusions in this matter can be obtained by studying finite pieces of trajectories.

Acknowledgments
This work was supported in part by the AGH University of Science and Technology, grant no. 11.11.120.343.

References

[1] Galias Z 2013 *IEEE Circuits and Systems Magazine* **13** 35–52
[2] Lozi R 2013 *Topology and Dynamics of Chaos* vol 84 ed Letellier C and Gilmore R (World Scientific) pp 29–64
[3] Hénon M 1976 *Communications in Mathematical Physics* **50** 69–77
[4] Galias Z and Tucker W 2014 *Chaos: An Interdisciplinary Journal of Nonlinear Science* **24** 013120 (14 pages)
[5] Galias Z and Tucker W 2015 *Chaos: An Interdisciplinary Journal of Nonlinear Science* **25** 033102 (12 pages)
[6] Biham O and Wenzel W 1989 *Physical Review Letters* **63** 819–822
[7] 2015 GNU MPFR library `http://www.mpfr.org`
[8] 2015 CAPD library `http://capd.ii.uj.edu.pl/`
[9] Moore R 1966 *Interval Analysis* (Englewood Cliffs, NJ: Prentice Hall)
[10] Neumaier A 1990 *Interval methods for systems of equations* (Cambridge University Press)
[11] Zgliczyński P 1997 *Nonlinearity* **10** 243–252
[12] Galias Z 2015 *Int. J. Bifurcation and Chaos* **25** 1550139 (14 pages)

Journal of Physics: Conference Series **692** (2016) 012002 doi:10.1088/1742-6596/692/1/012002

Border Collision and Smooth Bifurcations in a Family of Linear-Power Maps

Laura Gardini

DESP University of Urbino Carlo Bo, Urbino (Italy)

E-mail: laura.gardini@uniurb.it

Roya Makrooni

Faculty of Mathematical Sciences, Shahid Beheshti University, Tehran (Iran)

E-mail: r_makrooni@sbu.ac.ir

Abstract. In this work we describe some properties and bifurcations which occur in a family of linear-power maps typical in Nordmark' systems. The continuous case has been investigated by many authors since a few years, while the discontinuous case has been considered only recently. In particular, having a vertical asymptote, it gives rise to new kinds of bifurcations. Organizing centers related to codimension-two bifurcation points, due to the intersection of a border collision bifurcation and a smooth fold bifurcation of cycles having a different symbolic sequence are evidenced. It is shown the relevant role played by a codimension-two point existing on any border collision bifurcation curve, and related to the smooth fold bifurcation of cycles with the same symbolic sequence. We recall some of the properties proved up to now, evidencing the rich structure which is still to be understood.

1. Introduction

The study of piecewise smooth systems had a wide expansion in the last decade. This is related to the large number of applied models characterized by sharp switching between several states which are ultimately described by nonsmooth systems, continuous or discontinuous. Moreover, many applications in engineering may include nonlinearities in the map, as power functions. In this paper, we consider the well known one-dimensional piecewise smooth (PWS for short) map defined by two functions, $f_L(x)$ and $f_R(x)$, as follows:

$$x \to f_\mu(x) = \begin{cases} f_L(x) = ax + \mu & \text{if} \quad x \leq 0 \\ f_R(x) = bx^z + \mu & \text{if} \quad x > 0 \end{cases} \tag{1}$$

where a, b, z are real parameters and $\mu > 0$. The particular case $z = 1/2$ is the one more studied in the literature. It represents a continuous map, related to the square-root nonlinearity typical in Nordmark' systems and grazing bifurcations ([16], [17]). The piecewise linear case with $z = 1$ leads to the continuous skew tent map, whose dynamics are now well known (see e.g. [10], [24]). The power $z = 3/2$ was considered in [7] and in [6] where the normal-form mapping of sliding bifurcations is derived (leading to the map in (1) with power $z = 3/2$, $z = 2$ and $z = 3$, related to different cases of sliding bifurcations). Other examples of grazing and sliding bifurcations with nonlinear leading-order terms occur in power converters and in nonsmooth sliding-mode controls ([4], [5], [1]). System (1) with $z = 2$ is a particular case of the linear-logistic map considered in [22], [23]. A generalization of system (1) in the case $z > 0$ is also considered in

[2], by using the smooth function $f_R(x) = bx^z + cx + \mu$, which introduces a second critical point in the map.

In all the cases mentioned above, the power z takes positive values, which is related to a continuous PWS map, whose characteristic feature is the occurrence of border collision bifurcations (BCB for short), term which entered into use after the works by Nusse and Yorke ([18], [19]). In continuous PWS systems, the use of the skew-tent map as a normal form leads to a powerful analytical tool, which allows to determine the effect of the border collision of cycles of any period (applications can be found in [22], [23]). The same system in the case $z = 1/2$ but with different offsets (and thus a discontinuous system) was considered in [8], and the dynamics do not not differ significantly from the continuous case.

It is worth noticing that the power term z in (1) in the applied context appears through a Taylor series expansion of a nonlinear function, and so it can take only particular values (for example, in hard impact oscillators $z = 1/2$; in soft impact oscillators $z = 3/2$, and so on, as recalled above). However, also the case with real power $z < 0$ has been recently analyzed, and this leads to a discontinuous map, with a vertical asymptote, whose dynamic properties and bifurcations are very much different from those occurring in the continuous case. In discontinuous systems the classification of the possible different results of a BCB is still to be investigated, as well as the use of the PWL map as a normal form, and especially in maps with a vertical asymptote new phenomena arise, which are still to be understood. It was first considered in [20] where, besides the cases $z > 0$, the authors extend the analysis to the discontinuous case with $z < 0$. The particular case with $z = -1/2$ is also considered in [21]. However, the main results on the system in the case $z < 0$ have been shown in some recent papers ([11], [12], [14], [15], [13]), where many open problems are listed and left for further investigations.

The goal of the present work is to describe and evidence the bifurcations which are still to be investigated in this system, considering a generic real value for the power z, positive and negative. Recall that the main point in PWS systems, especially discontinuous, is that besides BCBs also standard bifurcations peculiar of smooth systems are involved. Thus, it is important to investigate the interactions between these two kinds of bifurcations. This is particularly relevant in the case of the system with an hyperbolic branch.

The existence of a vertical asymptote is not new in the engineering applications, and a peculiarity of such systems is that unbounded chaotic attractors can exist (considered also in [3]). However, to our knowledge, the cases known in the literature referring to unbounded chaotic sets are related to structurally unstable situations, and thus not related to true attractors (persistent under parameter perturbation). Differently, in this system with $z < 0$ it can be proved their occurrence and robustness ([11], [12]).

As already remarked, the main role in such systems is the occurrence of both BCBs and smooth bifurcations. However, in the noninvertible cases considered below, the smooth bifurcations are of different type: smooth flip bifurcations for $z > 0$ while smooth fold bifurcations for $z < 0$.

We recall that by using a change of variable we can get rid of one parameter, setting $\mu = 1$, which leads us to consider the following map:

$$x \to f(x) = \begin{cases} f_L(x) = ax + 1 & \text{if} \quad x \leq 0 \\ f_R(x) = bx^z + 1 & \text{if} \quad x > 0 \end{cases} \tag{2}$$

As above, the change of definition occurs at the discontinuity point $x = 0$, and as it is often used in PWS systems, the dynamical properties are studied making use of the symbolic notation based on the letters L and R corresponding to the two disjoint partitions

$$I_L = (-\infty, 0], \ I_R = (0, +\infty) \tag{3}$$

In the following section, as a prototype of the map $f(x)$ to describe the qualitative bifurcations occurring for $b < -1$ in the continuous system we consider examples with $z = 1/2$ and we shall recall some properties in Sec.2.1. While as a prototype of the map $f(x)$ to describe

Journal of Physics: Conference Series **692** (2016) 012002 doi:10.1088/1742-6596/692/1/012002

the qualitative bifurcations occurring in the discontinuous system with $z < 0$ we consider the case $z = -1/2$ and in Sec.2.2 we shall recall some properties and bifurcations, as well as open problems. In Sec.3 we show how rich is the bifurcation structure, and still to be understood, in the discontinuous case for $-1 < b < 0$ (when the correspondent continuous system has very simple dynamics).

2. Comparison between the cases $z > 0$ and $z < 0$, $b < -1$

In order to compare the two systems (continuous and discontinuous) let us first consider the parameter $b < -1$. The case with $-1 < b < 0$ will be commented in Sec.3, while $b > 0$ is not considered, as it leads to uninteresting (and trivial) dynamics both for $z > 0$ and $z < 0$.

For any value of z, in the region $b < -1$ we have that for $a \leq 1$ the dynamics cannot be divergent. The qualitative shape of the map is shown in Fig.1. For $z = 1/2$ the fixed point x_R^* on the R side is repelling, while for $z = -1/2$ no fixed point exists. In the continuous case there exists a bounded absorbing interval, $[f_R(1), 1]$, inside which the asymptotic dynamics are confined. In the discontinuous case there exists an unbounded absorbing interval $(-\infty, 1]$ and an unbounded chaotic set always exists (which may be of zero Lebesgue measure or of full measure).

Differently, for $a > 1$ the fixed point $x_L^* = -\frac{1}{a-1}$ on the L side exists and is repelling, and divergent trajectories exist. The immediate basin of ∞ is the interval $(-\infty, x_L^*)$ and its rank-1 preimage on the right side $f_R^{-1}((-\infty, x_L^*)) = (0, x_L^{*-1})$ where $x_L^{*-1} = \left(\frac{-a}{b(a-1)}\right)^{\frac{1}{z}}$. The total basin of divergent trajectories is clearly given by

$$\mathcal{B}_\infty = \bigcup_{k=0}^{\infty} f^{-k}((-\infty, x_L^*)) \tag{4}$$

Different dynamic behaviors occur depending on $x_L^{*-1} \lesseqgtr 1$ and in any case ($z \lessgtr 0$) when the parameters satisfy the condition $x_L^{*-1} = 1$ it is the homoclinic bifurcation of the fixed point x_L^*, and this leads to the bifurcation curve B_f of equation

$$B_f: \; b = -\frac{a}{a-1} \tag{5}$$

which is a kind of final bifurcation. It is clearly independent on the value of the parameter z, as it corresponds to $f_R \circ f_L(0) = x_L^*$, that is $f_R(1) = x_L^*$. However, the bifurcation leads to completely different dynamics in the two cases that we are comparing. In fact, for $z > 0$ the fixed point x_L^* is *not homoclinic* for parameter values above the curve B_f (and a bounded chaotic set exists in the absorbing interval $[f_R(1), 1]$). For parameter values on the curve B_f the fixed point becomes homoclinic (with all critical and degenerate homoclinic orbits, see [9]), and for parameter values below the curve B_f the fixed point x_L^* is homoclinic (with non critical and nondegenerate homoclinic orbits), so that an invariant set, chaotic repeller, exists, even if almost all the trajectories are divergent. See the different qualitative shapes of the map in the three different cases as drawn in Fig.1c, corresponding to the colored square points in the two-dimensional bifurcation diagram in Fig.1a.

A very different dynamic behavior occurs for $z < 0$, see the three different qualitative shapes of the map shown in Fig.1d (related to the square points in Fig.1b). The fixed point x_L^* is *not homoclinic* for parameter values below the curve B_f (and all the trajectories are divergent, except for the two points x_L^* and x_L^{*-1}), for parameter values on the curve B_f the fixed point becomes homoclinic (a unique homoclinic orbit exists, see the qualitative picture in red in Fig.1d, and all the other points have divergent trajectories, so that no invariant chaotic set can exist), and for parameter values above the curve B_f the fixed point x_L^* is homoclinic (with non critical and nondegenerate homoclinic orbits), so that an invariant chaotic repeller exists in the interval $]x_L^*, 1[$ and the basin of divergent trajectories has a positive measure (even if the attractor at infinity may coexist with an attracting cycle in the interval $]x_L^*, 1[$).

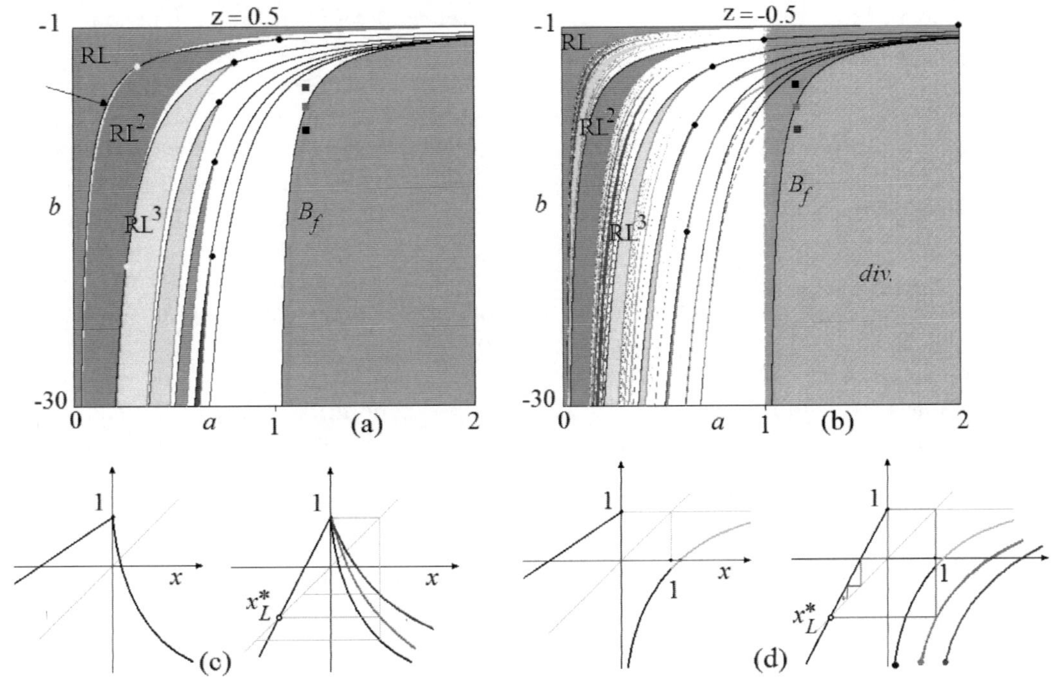

Figure 1. Two-dimensional bifurcation diagrams in the (a, b)-parameter plane, at $z = 0.5$ in (a), $z = -0.5$ in (b). In Fig.1c,d qualitative shape of the map is shown for $z > 0$ in (c) and for $z < 0$ in (d).

As we can see from Fig.1a,b, the main differences between the continuous and discontinuous cases occur in the range $0 < a < 1$. The equation of the BCB curves of the basic cycle with symbolic sequences RL^n comes from the equation

$$f_L^n \circ f_R(1) = 1 \tag{6}$$

and thus it is independent on the parameter z. It is possible to write the equation in explicit form:

$$B_{RL^n}: \ b = -\frac{1 - a^n}{a^{n-1}(1 - a)} \tag{7}$$

but the dynamic properties crossing the bifurcation curve B_{RL^n} are different depending on z.

2.1. Crossing a BCB curve B_{RL^n}, $z > 0$

In the continuous case ($z > 0$), as it occurs in the skew-tent map (see [?]), the crossing of a bifurcation curve B_{RL^n} corresponds to a fold border collision bifurcation, leading (as the parameter b is decreased) to a pair of cycles, one with symbolic sequence RL^n (which may be attracting or repelling) and a repelling one with symbolic sequence R^2L^{n-1}. If the cycle RL^n is attracting then decreasing the parameter b it becomes repelling via a smooth flip bifurcation. The flip bifurcation curves are given by:

$$\Psi_{RL^n}: \ b = -\frac{1}{za^n}\left(\frac{1 - a^{n+1}}{1 - a}\frac{z}{z+1}\right)^{1-z} \quad , \quad z = \frac{1}{2} \text{ in our example} \tag{8}$$

Each flip bifurcation curve Ψ_{RL^n} intersects the BCB curve B_{RL^n} in a codimension-two point (a_n^ψ, b_n^ψ), $n \geq 1$, which separates the two behaviors: crossing B_{RL^n} at $a < a_n^\psi$ the basic cycle RL^n, decreasing b, appears attracting, while crossing B_{RL^n} at $a \geq a_n^\psi$ also the basic cycle RL^n appears repelling (that is, the border collision is related to a pair of repelling cycles, the flip bifurcation does not occur and the fold bifurcation leads to the existence also of all the subharmonics, as unstable cycles). In Fig.1a some codimension-two points (a_n^ψ, b_n^ψ) are evidenced by black circles.

As it is typical in the skew-tent map, the basic cycles RL^n are the only ones which may be attracting, and after their supercritical flip bifurcation chaotic attractors in bands are expected: $2(n+1)-$pieces, $(n+1)-$ pieces and $1-$piece, respectively, decreasing b (the qualitative changes occur at the homoclinic bifurcations of the basic cycles appeared at the fold BCB), up to the final bifurcation curve B_f.

However, it is worth to note the difference existing in the continuous map under consideration with respect to the skew-tent map. That is, while in the skew-tent map bistability is not allowed (the flip bifurcations are always supercritical, we cannot have two coexisting attracting cycle, or an attracting cycle and a chaotic attractor), in the present case we can have bistability, even if the map is unimodal. This is related to the fact that the flip bifurcation of a basic cycle may be subcritical. Indeed, we can evidence other codimension-two points $(a_n^{\psi-}, b_n^{\psi-})$, on the BCB curves, due to the intersection, for any $n \geq 2$, of the BCB curve B_{RL^n} with the flip bifurcation curve $\Psi_{RL^{n-1}}$ (see the yellow points in Fig.1a, corresponding to $n = 2$ and $n = 3$, $B_{RL^2} \cap \Psi_{RL}$ and $B_{RL^3} \cap \Psi_{RL^2}$, respectively). The role of such a codimension-two point is the following: for $a < a_n^{\psi-}$ the stability regions of the basic cycles RL^n and RL^{n-1} are overlapping: there exists a region in the $(a, b)-$parameter plane related to their coexistence (in Fig.1a an arrow indicates a yellow strip which is the overlapping region between the stability region of RL and RL^2). For the cycle RL^{n-1} we have that crossing the flip bifurcation curve $\Psi_{RL^{n-1}}$ above the codimension-two point (for $a > a_n^{\psi-}$), decreasing b, *the flip bifurcation is supercritical* while crossing it below the codimension-two point (for $a < a_n^{\psi-}$) *the flip bifurcation is subcritical.*

This implies that for $a < a_n^{\psi-}$ a repelling cycle with symbolic sequence $RL^{n-1}RL^{n-1}$ must have been appeared before (at a larger value of b). Indeed, at the crossing of the BCB curve B_{RL^n} two attractors exist, and thus a pair (by continuity) of repelling cycles must appear before such border collision, via a fold BCB of the map $f^{2(n-1)}$. The basin of attraction of the attracting cycle RL^{n-1} is bounded by the unstable cycle with symbolic sequence $RL^{n-1}RL^{n-1}$, and the basin shrinks to the cycle itself at the subcritical flip bifurcation.

2.2. Crossing a BCB curve for $z < 0$

Turning to the discontinuous case $z < 0$, the crossing of the BCB curve B_{RL^n} given in (7) is now related to a unique basic cycle with symbolic sequence RL^n. Differently from above, it is now suitable to comment the effect of the bifurcations as the parameter b is increased. Increasing b, the crossing of the BCB curve B_{RL^n} leads to the disappearance of a single cycle RL^n which may be attracting or repelling. The difference is related to the crossing (at smaller values of b) of a fold bifurcation curve or not. In fact, in the discontinuous case, since

$$f_R'(x) = \frac{bz}{x^{1-z}} > 0 \ , \ f_R''(x) = -\frac{zb(1-z)}{x^{2-z}} < 0 \tag{9}$$

we have that the slope of map f in (2) (and of any iterate f^n) is positive in all the points, thus no flip bifurcation of cycles can occur. Instead, smooth fold bifurcations can occur. Smooth fold bifurcations of basic cycles RL^n can be detected considering the composite function $F_{RL^n}(x) = f_L^n \circ f_R(x)$ and its tangency with the diagonal in a point $x^*_{RL^n} \in (0, 1]$. The explicit equation of the fold bifurcation curve Φ_{RL^n} is as follows:

$$\Phi_{RL^n} : \ b = +\frac{1}{za^n}\left(\frac{1-a^{n+1}}{1-a}\frac{z}{z-1}\right)^{1-z} \quad , \quad z = -\frac{1}{2} \text{ in our example} \tag{10}$$

The fold bifurcation curve Φ_{RL^n} intersects the BCB curve B_{RL^n} at a codimension-two point (a_n^*, b_n^*), where a_n^* is the solution of the equation $a\frac{1-a^n}{1-a} = -\frac{1}{z}$ and $b_n^* = \frac{1}{za_n^*}$. For any $n \geq 1$ they satisfy the inequalities

$$a_\infty = \frac{1}{1-z} < a_{n+1}^* < a_n^* < a_1^* = -\frac{1}{z} \tag{11}$$

In Fig.1b some codimension-two points (a_n^*, b_n^*) are evidenced by black circles. The role of a codimension-two point (a_n^*, b_n^*) is the following: crossing a BCB curve B_{RL^n} above its codimension-two point (a_n^*, b_n^*), i.e. for $a \geq a_n^*$, the fold bifurcation does not occur (it is virtual) and increasing b, at the BCB curve a unique repelling cycle with symbolic sequence RL^n appears and persist for any larger value of b. While for $a < a_n^*$, the fold bifurcation occurs before the border collision, leading to the existence of a pair of cycles with symbolic sequence RL^n, one attracting and one repelling. Increasing b the attracting one disappears by border collision crossing the curve B_{RL^n}, leaving the repelling cycle with the same symbolic sequence, existing for any larger value of b.

Now coexistence of attracting cycle cannot occur, while infinitely many families of attracting cycles exist, not only those related to basic cycles. Periodicity regions associated with many cycles can be seen in Fig.1b.

In Fig.2 we show an enlargement, and crossing a vertical segment from the BCB of an attracting cycle RL^3 to the BCB of an attracting cycle RL^2 we can prove the existence of infinitely many families of cycles.

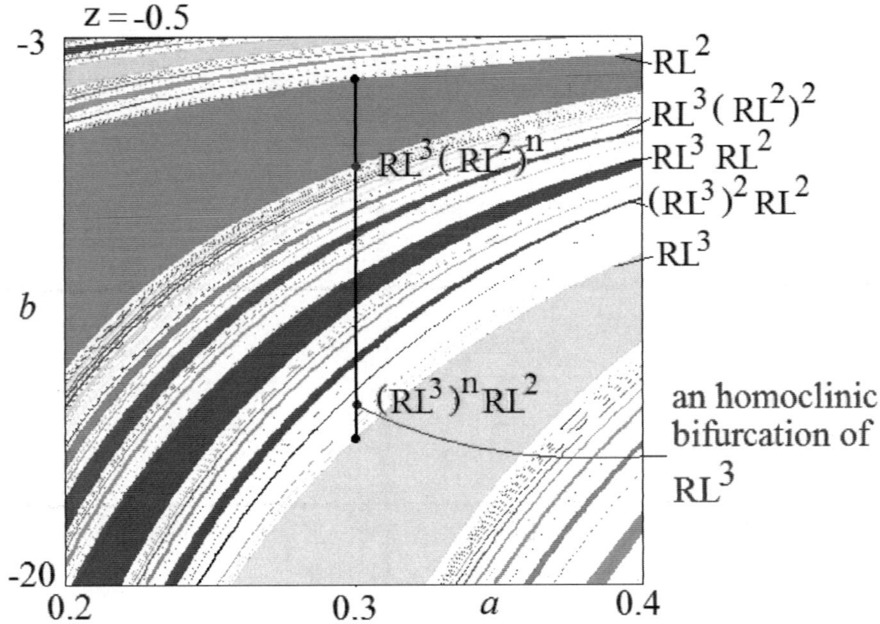

Figure 2. Two-dimensional bifurcation diagrams in the (a, b)-parameter plane, enlargement of a portion of Fig.1b.

Similarly to the description given above for the basic cycles, also for the other families we can prove the existence of the BCB curves, and the cycle which undergoes the collision may be attracting or repelling, depending on the previous crossing of a fold bifurcation curve or not.

In the discontinuous case, the dynamics of map f can be described by using the first return map $F_r(x)$ in the interval $I = [0, 1]$. As shown in [14], $F_r(x)$ is a discontinuous map with infinitely many branches defined as follows:

$$F_r(x) := \begin{cases} F_{RL^n}(x) = f_L^n \circ f_R(x) & \text{if } \xi_{n+1} \leq x \leq 1 \\ F_{RL^{n+1}}(x) = f_L^{n+1} \circ f_R(x) & \text{if } \xi_{n+2} \leq x < \xi_{n+1} \\ \vdots & \vdots \\ F_{RL^{n+j}}(x) = f_L^{n+j} \circ f_R(x) & \text{if } \xi_{n+j+1} \leq x < \xi_{n+j} \\ \vdots & \vdots \end{cases} \quad (12)$$

where $n \geq 0$ is the smallest integer for which

$$f_L^n \circ f_R(1) \in [0,1) \tag{13}$$

with

$$F_{RL^m}(x) = a^m b x^z + \frac{1-a^{m+1}}{1-a} \tag{14}$$

and the discontinuity points are preimages of the origin given by

$$\xi_{m+1} = f_R^{-1} \circ f_L^{-m}(0) = \left(\frac{-b}{\frac{a^m-1}{a^m(a-1)}+1} \right)^{-\frac{1}{z}} \tag{15}$$

which have as limit value, for $m \to \infty$, the point $x = 0$.

For any $m \geq n+1$, $F_{RL^m}(\xi_{m+1}) = 0$ and $F_{RL^m}(\xi_m) = 1$ hold, while the rightmost branch satisfies $F_{RL^n}(\xi_{n+1}) = 0$ and $F_{RL^n}(x) \in (0,1)$ for $\xi_{n+1} < x < 1$. The case

$$f_L^n \circ f_R(1) = 0 \ \ (\ f_L^{n+1} \circ f_R(1) = 1 \ ) \tag{16}$$

corresponds to the BCB of a basic cycle with symbolic sequence RL^{n+1}.

An example of first return map is shown in Fig.3

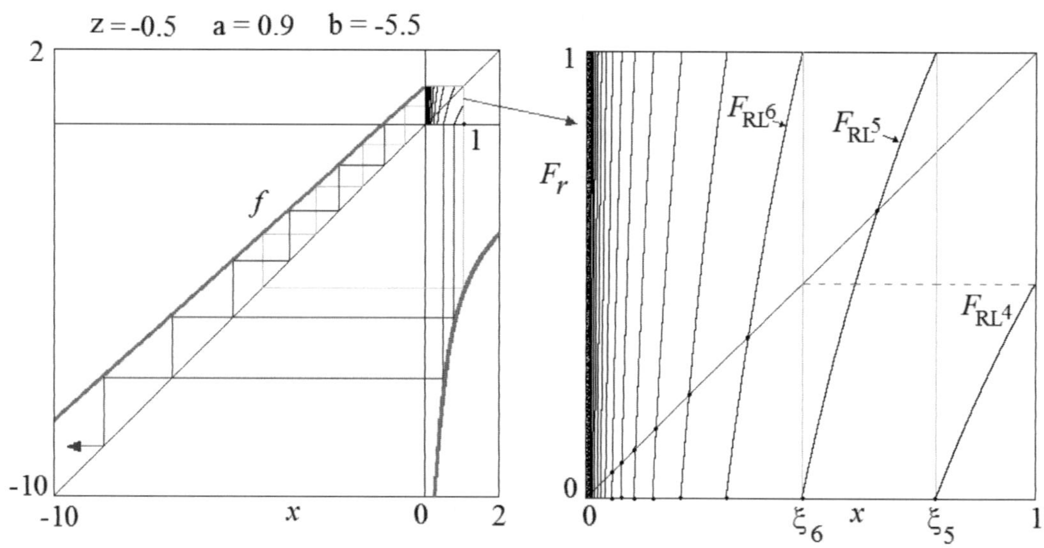

Figure 3. Graph of map f and first return map $F_r(x)$, shown also enlarged.

Only the rightmost branch RL^m can have a fixed point with slope smaller than 1. When this does not occur, and the slope of the first return map is larger than 1 in all the points of continuity, as in the example shown in Fig.3, then the first return map is chaotic in the whole interval $[0,1]$, which means that map f is chaotic in the whole unbounded interval $(-\infty,1]$, as proved in [12].

Increasing b the value in $x=1$ increases and the diagonal is reached at the BCB value of the cycle RL^m (which appears, unstable, and persists for larger values of b). This is the scenario related to the crossing of a BCB after its codimension-two point, and the robustness of the unbounded chaotic attractor $(-\infty,1]$ is proved for any $a > a_n^*$ and $b < b_n^*$ ([12], [14]).

The dynamic behavior occurring when a BCB is related to a fold bifurcation is illustrated in Fig.4.

When the rightmost branch RL^m has points with slope smaller than 1, then increasing b the fold bifurcation will occur leading to a pair of cycles, one attracting and one repelling (with the

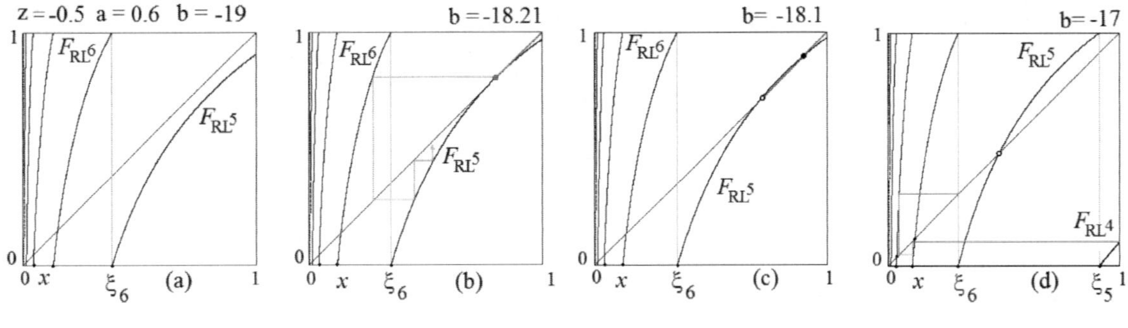

Figure 4. Graph of the first return map in the case related to a fold bifurcation.

same symbolic sequence RL^m). When an attracting cycle exists almost all the points belong to its basin of attraction, but its boundary includes an invariant chaotic set which accumulates to $x = 0$, which means that map f has an unbounded chaotic set in $(-\infty, 1]$. Increasing b the rightmost point of the attracting cycle approaches $x = 1$, merging with it at its BCB. The attracting cycle disappears leaving the repelling cycle, which exists for any larger value of b.

It is easy to prove that:

- all the (infinitely many) repelling fixed points of the first return map $F_r(x)$ are homoclinic, and f has always an unbounded chaotic set in the interval $(-\infty, 1]$;

- let $F_{RL^m}(x)$ be the rightmost branch of $F_r(x)$. If $F'_{RL^m}(1) = ba^m z \geq 1$ then the interval $(-\infty, 1]$ is an unbounded chaotic attractor of f.

Up to now we have considered only the BCB of basic cycles B_{RL^n}, which do not depend on the parameter z, but the BCB of cycles of different symbolic sequence also depend on z. In fact, any BCB involving a not basic cycle of f is represented, for $j \geq (m+1)$ by the condition

$$F_r(1) = \xi_j^{-k} \quad (as \ \ F_r^{k+1}(1) = \xi_j \ , \ F_{RL^j} \circ F_r^{k+1}(1) = 1) \tag{17}$$

where $F_{RL^m}(x)$ is the rightmost branch of $F_r(x)$) and $k \geq 0$, leading to the BCB of a cycle of the first return map F_r of period $(k+2)$, and a cycle of f with symbolic sequence related to the involved branches of F_r, say

$$RL^m RL^{n_1} ... RL^{n_k} RL^j \tag{18}$$

On each BCB curve we can reason as for the BCB of basic cycles, determining a related codimension-two point. Let D be the first derivative of the function $F_{RL^j} \circ F_r^{k+1}(x)$ in the point $x = 1$. Then the BCB leads (increasing b) to the appearance of a repelling cycle or to the disappearance of an attracting cycle depending on $D \geq 1$ or $D < 1$, respectively. The case $D = 1$ corresponds to the codimension-two point at which the BCB occurs simultaneously with a fold bifurcation of cycles having the same symbolic sequence.

Although all the repelling cycles of the first return map $F_r(x)$ are homoclinic, a new homoclinic explosion occurs whenever

$$F_r^k(1) = x^* \tag{19}$$

for some $k \geq 0$, where x^* is a point of a repelling cycle of F_r.

For example, in Fig.5a we have $F_r(1) = x_{RL^3}$ where x_{RL^3} is the repelling fixed point of the branch F_{RL^3}, thus for a larger value of b, when $F_r(1) > x_{RL^3}$ holds, there is a new preimage, $F^{-1}_{RL^2}(x_{RL^3})$, which did not exist before (for $F_r(1) < x_{RL^3}$), from which an explosion of infinitely many new homoclinic orbits of x_{RL^3} can be detected.

So we can state that for any BCB of map f, whichever is the symbolic sequence of the colliding cycle, the structure is qualitatively represented in Fig.6, where the codimension-two point (related to the derivative $D = 1$) separates the two different kinds of collision: after that

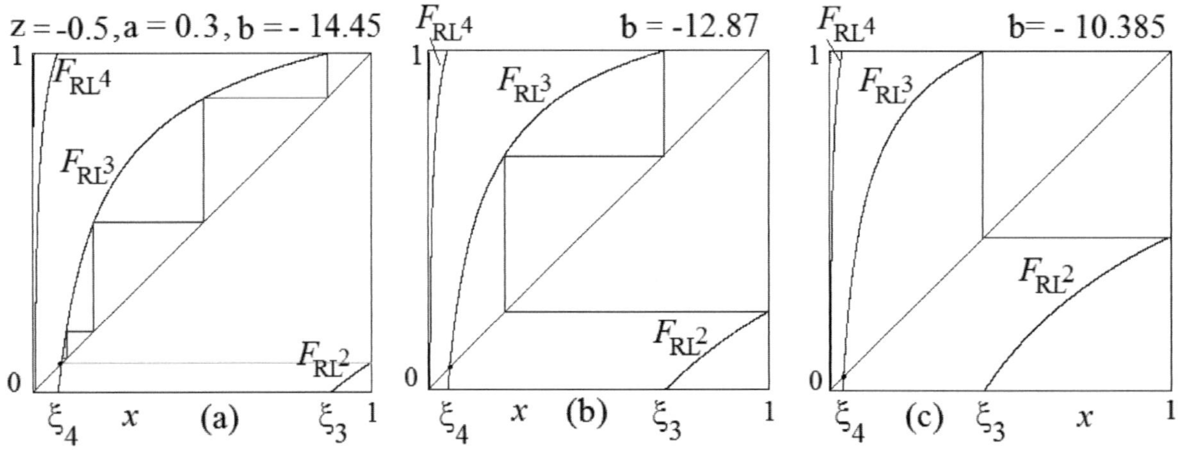

Figure 5. Graph of the first return map in the case related to an homoclinic bifurcation in (a), and to a BCB in (b) and (c).

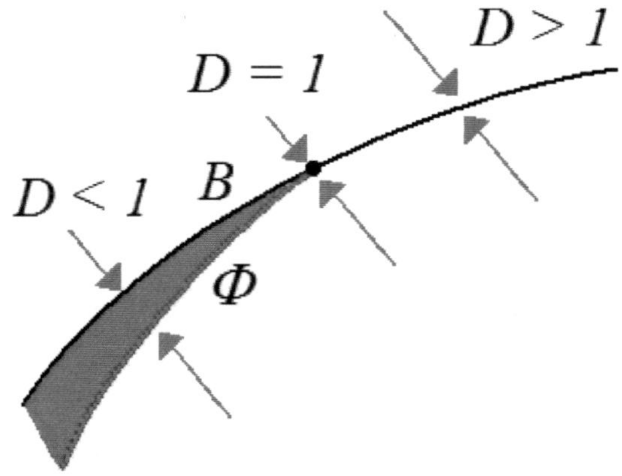

Figure 6. Qualitative representation of a BCB curve and related codimension-two point on it.

point, crossing the curve increasing b a repelling cycle appears, before that point crossing the BCB curve an attracting cycle disappears, leaving the unstable one with the same symbolic sequence, which means that at a smaller value of b the a fold bifurcation must have been occurred. Moreover:

- any BCB curve is a limit set of infinitely many BCB curves, on both sides when related to a repelling cycle, only from above when related to an attracting cycle;
- each fold bifurcation curve is a limit set of infinitely many homoclinic bifurcation curves and BCB curves, only from below:
- each homoclinic bifurcation curve is a limit set of infinitely many BCB curves, both from below and above;
- chaotic bands do not occur.

Referring to the vertical segment in Fig.2, at $a = 0.3$ for $-15.4 = b(B_{RL^3}) < b < b(B_{RL^2}) = -4.3$, let us show that the two families of BCB curves related to cycles with the symbolic sequence obtained by concatenation of RL^2 and RL^3 must exist, that is cycles $RL^2(RL^3)^n$ and $(RL^2)^n RL^3$ for any $n \geq 1$.

Considering the shape of the first return map in Fig.5, a sequence of preimages of the discontinuity point ξ_3 with the branch F_{RL^3} is accumulating to the repelling fixed point x_{RL^3} and after the homoclinic bifurcation shown in Fig.5a, increasing b, the point $F_r(1) = F_{RL^2}(1)$ increases (and it reaches the value 1 at the BCB B_{RL^2}), so that there must exist values of b, say b_k for any $k \geq 0$, such that at $b = b_k$ the BCB curves having the following equations are crossed:

$$F_{RL^3}^{k+1} \circ F_{RL^2}(1) = 1, \ \ k \geq 0$$

which are the BCB curves of cycles of f having symbolic sequence $RL^2(RL^3)^{k+1}$ for any $k \geq 0$. Since for $k \to \infty$ the values b_k tend to $b(x_{RL^3})$ these border collisions have as limit set the homoclinic bifurcation value of this cycle (occurring for $F_{RL^2}(1) = x_{RL^3}$). In Fig.5b,c the last bifurcations related to $k = 1$ and $k = 0$, respectively, are shown.

For the second family, notice that after the last bifurcation commented above, for $k = 0$ (which corresponds to $F_{RL^2}(1) = \xi_3$), we can consider the preimages of the discontinuity point ξ_3 also with the branch F_{RL^2}, as shown in Fig.7. It is clear that for any $n > 1$ the condition

$$F_{RL^2}^n(1) = \xi_3 \tag{20}$$

must occur, leading (from $\xi_3 = F_{RL^3}^{-1}(1)$) to the BCB curves of equation $F_{RL^3} \circ F_{RL^2}^n(1) = 1$, $n \geq 2$ which have as limit set the fold bifurcation curve Φ_{RL^2}.

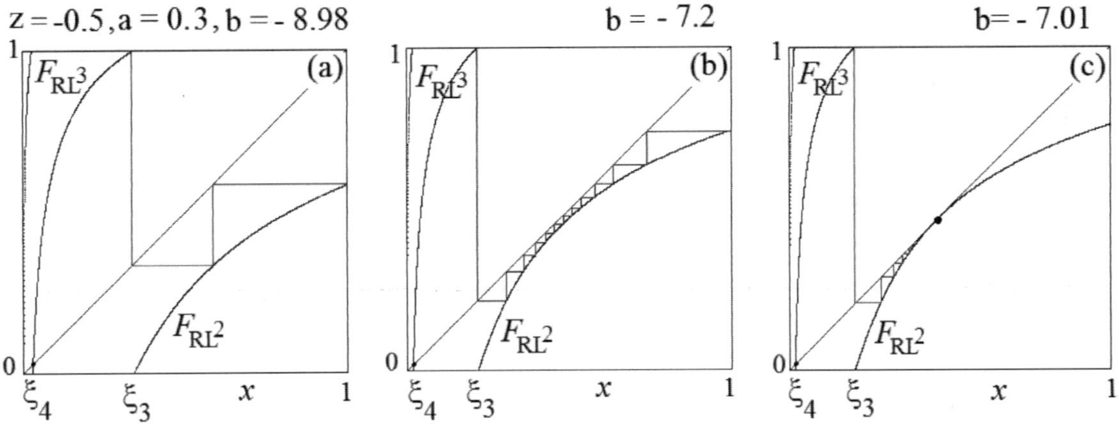

Figure 7. Graph of the first return map in the case related to some BCBs.

For each BCB we have a qualitative shape as that shown in Fig.6. In the considered example in Fig.2 we cannot say precisely which of them are crossed with attracting or repelling cycles for $a = 0.3$. Numerically, it seems that the second family of cycles with symbolic sequence $(RL^2)^n RL^3$ for any $n \geq 1$ is associated with the BCB of attracting cycles (and thus also the related fold bifurcation curves are crossed).

The two families described above are just a simple example. It is clear that many more cycles undergo a BCB in the vertical segment at $a = 0.3$ shown in Fig.2, and especially the structure existing between the lower BCB bifurcation of the cycle RL^3 and the homoclinic bifurcation of x_{RL^3} commented above is still to be understood.

3. Comparison for $-1 < b < 0$

The case with $-1 < b < 0$ is not interesting in the continuous system ($z > 0$), while the discontinuous system is very much interesting and new. In Fig.8a we show the whole range $b < 0$ by using, in the vertical axis, the scaled value $S(b) = arctan(b)$ in the interval $[-\frac{\pi}{2}, 0]$. It can be seen that all the bifurcation curves discussed in the previous section for $b < -1$ are issuing from the point $(a, b) = (0, -\infty)$ which behaves as organizing center. The qualitative shape of the right branch $f_R(x)$ is shown, and increasing b the BCB of the basic cycles R^nL occur for any $n \geq 1$ having as limit set (for $n \to \infty$) the fold bifurcation curve of $f_R(x)$, Φ_R, obtained from (10) for $n = 0$:

$$\Phi_R : \ b = \frac{1}{z}\left(\frac{z}{z-1}\right)^{1-z}, \ \ z = -\frac{1}{2} \text{ in our example} \tag{21}$$

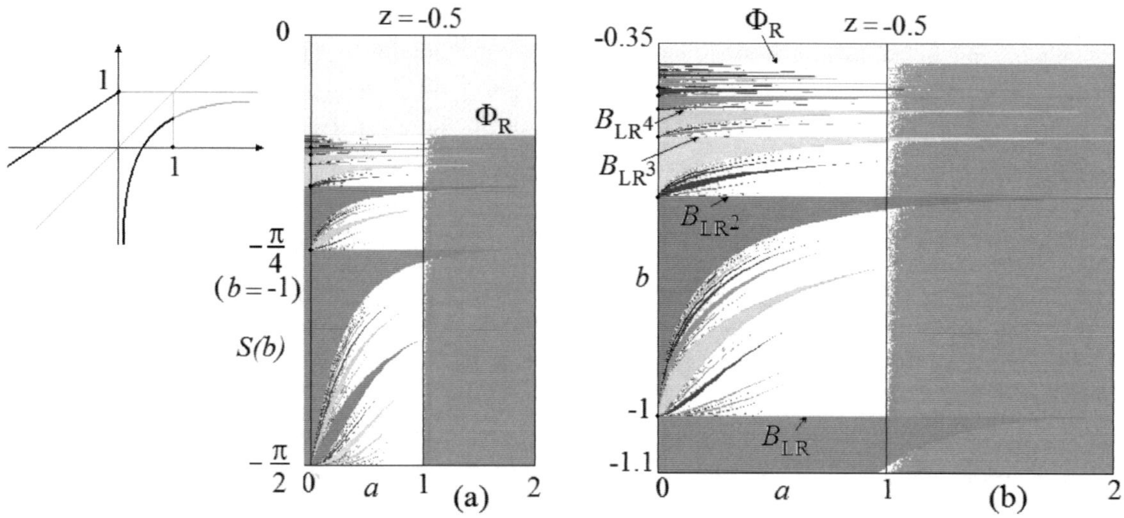

Figure 8. In (a) two-dimensional bifurcation diagram in the $(a, S(b))$-plane where $S(b) = arctan(b)$. In (b) two-dimensional bifurcation diagram enlarged in the (a, b)-plane.

What is evident from Fig.8b (an enlarged portion in the (a, b)-parameter plane) is that besides the organizing center in $(a, b) = (0, -\infty)$, there are infinitely many organizing centers related to codimension-two points which comes from the intersection of bifurcation curves of cycles having different symbolic sequences, that is $(0, b_n^B) = B_{R^nL} \cap \Phi_{R^{n+1}L}$. The organizing centers belong to the line $a = 0$, at which the shape of the map is particular, having a flat branch on the left side (slope 0). In such cases, increasing the parameter b it is known that a pure period incrementing structure (with increment 1) takes place. Moreover, as shown in [13], the vertical asymptote leads to a peculiarity: each border collision is always related to a pair of cycles, one bounded and one, with period +1, having a periodic point at infinity, as shown in the one-dimensional bifurcation diagram in Fig.9 with $S(x) = arctan(x)$ on the vertical axis.

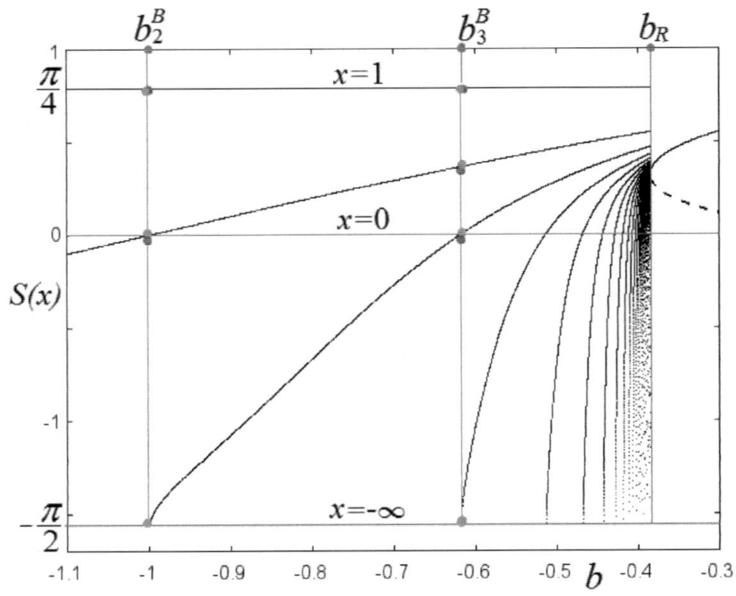

Figure 9. One-dimensional bifurcation diagram at $a = 0$, showing $S(x)$ as a function of b, where $S(x) = arctan(x)$. In red the points of the cycle of period n at its BCB, in green the points of the cycle of period $(n + 1)$, with one point at infinity.

In [15] it is shown that each bifurcation value of B_{R^nL} is independent on a, thus $b = b_n^B$ are the equations of the BCB curves B_{R^nL} (horizontal straight lines in Fig.8). In that work it is shown that also in this case the properties of the map can be studied making use of the return map in the in terval $[0, 1]$, and that the BCB curves have the same property as above, qualitatively expressed in Fig.6

The main point is that the rich structure discussed above for the case $b < -1$ (and related to curves issuing from $(0, -\infty)$) can be repeated with obvious changes from any organizing center $(0, b_n^B)$, and the related structure is still an open problem.

4. Conclusions

In this work we have considered the map f in (2) with power z both for positive and negative values, emphasizing the differences in the bifurcation structures in the (a, b) parameter plane. In particular we have shown how rich is the bifurcation structure in the case $z < 0$, leading to many open problems. We have shown that a a suitable tool for its study is the first return map $F_r(x)$ in the interval $[0, 1]$. We are confident that the existing bifurcation structure may be explained by using the map $F_r(x)$, both in the case $b < -1$ and $-1 < b < 0$. The interactions between two kinds of bifurcations (BCBs and standard smooth fold bifurcations) has been shown, and on each BCB curve (related to any possible cycle) it appears related to a specific codimension-two point which can be determined by the condition $D = 1$ as shown in Sec. 2.2. Moreover, a nice property of the system in the case $z < 0$ is that the interval $(-\infty, 1]$ is an unbounded chaotic attractor of f which is structurally stable in a wide portion of the parameter plane for $b < -1$.

Acknowledgements

The work of L. Gardini has been performed under the auspices of the COST action IS1104 "The EU in the new economic complex geography: models, tools and policy evaluation", www.gecomplexitycost.eu, and within the activities of the GNFM (National Group of Mathematical Physics, INDAM Italian Research Group). R. Makrooni thanks the Department DESP of the University of Urbino for the hospitality during her six months of stay in Italy.

Journal of Physics: Conference Series **692** (2016) 012002 doi:10.1088/1742-6596/692/1/012002

References

[1] Angulo F di Bernardo M Fossas E and Olivar G 2005 *IEEE Transactions on Circuits and Systems-I*, vol 52 p 366-378
[2] Avrutin V Dutta P S Schanz M and Banerjee S 2010 *Nonlinearity* vol 23 p 445–463
[3] Bischi G I Mira C and Gardini L 2000 *Int. Journal of Bifurcation and Chaos* vol 10 p 1437-1469
[4] di Bernardo M Budd C J and Champneys A R 1998 *Nonlinearity* vol 11 p 858-890
[5] di Bernardo M Budd C J and Champneys A R 2001 *Physica D* vol 154 p 171-194
[6] di Bernardo M Kowalczyk P and Nordmark A B 2002 *Physica D* vol 170 p 170-175
[7] Dankowicz H and Nordmark A B 2000 *Physica D* vol 136 280-302
[8] Dutta P S and Banerjee S 2010 *Discrete and Continuyous Dynamical System, Series B* vol 14 p 961–976
[9] Gardini L Sushko I Avrutin V and Schanz M 2011 *Chaos Solitons & Fract.* vol 44 p 33–49
[10] Maistrenko Y L Maistrenko V L Vikul S I 1998 *Chaos Solitons & Fract.* vol 9 p 67-75
[11] Makrooni R and Gardini L 2015 Discussion Paper N.7 ISSN: 2409-7497, http://econpapers.repec.org/paper/cstwpaper/
[12] Makrooni R Abbasi N Pourbarat M and Gardini L 2015 *Chaos, Solitons & Fractals* vol 77 p 310–318
[13] Makrooni R Gardini L and Sushko I 2015 *Int. Journal of Bifurcation and Chaos* to appear
[14] Makrooni R Khellat F and Gardini L 2015 *Journal of Difference Equations and Applications* DOI:10.1080/10236198.2015.1045893
[15] Makrooni R Khellat F and Gardini L 2015 *Journal of Difference Equations and Applications* DOI: 10.1080/10236198.2015.1046855
[16] Nordmark A B 1991 *Journal of Sound and Vibration* vol 145 p 279–297
[17] Nordmark A B 1997 *Physical Review E* vol 55 p 266-270
[18] Nusse H E and Yorke J A 1992 *Physica D* vol 57 p 39-57
[19] Nusse H E and Yorke J A 1995 *Int. Journal of Bifurcation and Chaos* vol 5 p 189–207
[20] Qin Z Yang J Banerjee S and Jiang G 2011 *Discrete and Continuous Dynamical System, Series B* vol 16 p 547-567
[21] Qin Z Zhao Y and Yang J 2012 *Int. Journal of Bifurcation and Chaos* vol 22 p 1250112
[22] Sushko I Agliari A and Gardini L 2005 *Discrete and Continuous Dynamical Systems, Serie B* vol 5 p 881-897
[23] Sushko I Agliari A and Gardini L 2006 *Chaos Solitons & Fractals* vol 29 p 756-770
[24] Sushko I and Gardini L 2010 *Int. Journal of Bifurcation and Chaos* vol 20 p 2045–2070

Disrupted bandcount doubling in an AC-DC boost PFC circuit modeled by a time varying map

Viktor Avrutin

Institute for Systems Theory and Automatic Control, University of Stuttgart, Stuttgart, Germany

E-mail: viktor.avrutin@ist.uni-stuttgart.de

Zhanybai T. Zhusubaliyev

Department of Computer Science, Southwest State University, Kursk, Russia

E-mail: zhanybai@hotmail.com

Abdelali El Aroudi

Department of Electronics, Electrical Engineering, Automatic Control, Universitat Rovira i Virgili, Tarragona, Spain

E-mail: abdelali.elaroudi@urv.cat

Danièle Fournier-Prunaret

INSA, LAAS-CNRS, Université de Toulouse, Toulouse, France

E-mail: daniele.fournier@insa-toulouse.fr

Germain Garcia

INSA, LAAS-CNRS, Université de Toulouse, Toulouse, France

E-mail: garcia@laas.fr

Erik Mosekilde

Department of Physics, The Technical University of Denmark, Lyngby, Denmark

E-mail: erik.mosekilde@fysik.dtu.dk

Abstract. Power factor correction converters are used in many applications as AC-DC power supplies aiming at maintaining a near unity power factor. Systems of this type are known to exhibit nonlinear phenomena such as sub-harmonic oscillations and chaotic regimes that cannot be described by traditional averaged models. In this paper, we derive a time varying discrete-time map modeling the behavior of a power factor correction AC-DC boost converter. This map is derived in closed-form and is able to faithfully reproduce the system behavior under realistic conditions. In the chaotic regime the map exhibits a sequence of bifurcation similar to a bandcount doubling cascade on the low frequency. However, the observed scenario appears in some sense incomplete, with some gaps in the bifurcation diagram, whose appearance to our knowledge has never been reported before. We show that these gaps are caused by high frequency oscillations.

Journal of Physics: Conference Series **692** (2016) 012003 doi:10.1088/1742-6596/692/1/012003

1. Introduction

Power factor correction (PFC) circuits are power supplies that regulate an output voltage while providing a near unity power factor (PF) in the sense that the input current is proportional to the input voltage [1, 2]. Although, a variety of circuit topologies and control methods have been developed for PFC applications, the boost converter working in continuous conduction mode is commonly chosen for medium and high power (above several hundred W) applications because of its low conduction losses and reduced electromagnetic interferences filtering requirements [2].

The output voltage of a boost PFC converter (the bus voltage) must be always higher than the peak line voltage for an appropriate functioning of the system. For usual line input voltage applications (85-265 V RMS), the bus voltage is usually fixed in the range 380-400 V DC, which is achieved by an appropriate external voltage feedback compensating loop.

While the output voltage of an AC-DC PFC circuit is regulated to a desired DC value, its input is a sinusoidal time varying signal. This leads the duty cycle to vary between one and a minimum value. One of the most important tasks in the design of PFC power supplies is the implementation of the inner control loop.

While the main design criterion of maintaining a near unity PF can be achieved, in average, by using traditional control design techniques based on averaged modeling, it is well known that stability of the fundamental periodic orbit can not be achieved for all parameter values. Recently it has been shown both by numerical simulations and by experimental measurements that boost PFC converter can exhibit period doubling cascades and sub-harmonic oscillations at the line frequency as well as bubbling at the switching frequency [3, 4, 5]. This is a very undesirable behavior as it can jeopardize considerably the system performances by increasing its total harmonic distortion, as confirmed both experimentally and numerically (see for example recent publications [6, 7]). Realizing the practical importance of PFC AC-DC converters, a series of studies dealing with the nonlinear behavior in these systems have been carried out on this topic [8, 9, 10, 11].

In fact, the occurrence of bifurcations in power electronics circuits was reported in the pioneer works of Hamill and co-workers [12, 13] where the importance of studying the dynamical behavior of these systems by using iterative mappings was pointed out. Among the earliest works providing a detailed numerical and analytic investigation of bifurcations and transitions to chaos in DC-DC converters it is worth mentioning the paper by Baushev and Zhusubaliyev [14]. In this work it is shown that such systems can possess a wide range of parameters with many locally stable limit cycles with different dynamic characteristics, including regions of multistability. These cycles arise in hard transitions, and with changing parameters each of them can undergo either a finite or an infinite sequence of period-doubling bifurcations, resulting in the transition to chaos.

While in DC-DC converters, the map describing the dynamical behavior is time invariant (autonomous) [12, 13, 14], in the case DC-AC converters [15] and AC-DC converters, the map takes a time varying (non-autonomous) form [8, 16, 17]. The goal of the present paper is to present a simple time-varying map modeling the behavior of an AC-DC PFC boost converter, and to report an unusual bifurcation phenomenon which, to our knowledge, has never been reported before.

The paper is organized as follows. First, in Secs. 2 and 3 we describe the considered circuit, its model in continuous time and the discrete-time model resulting from it. Next, in Sec. 4 we show the examples of simulations in continuous and discrete time domains. Thereafter, in Sec. 5 we discuss bifurcation structures formed by chaotic attractors in the considered system. The bifurcation diagrams obtained under variation of one of the parameters appear surprisingly incomplete. We show that this is not a problem related to numerical inaccuracy but a phenomenon caused by high frequency oscillations.

Journal of Physics: Conference Series **692** (2016) 012003 doi:10.1088/1742-6596/692/1/012003

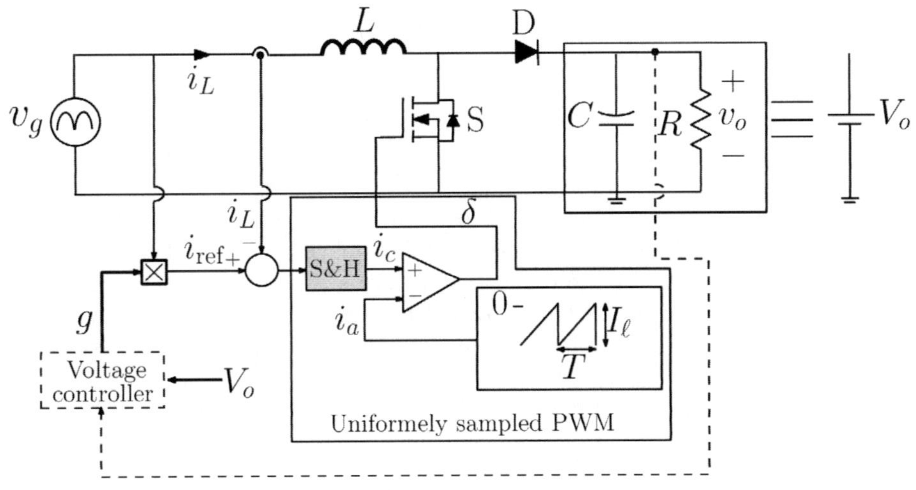

Figure 1. Boost PFC circuit designed to provide a regulated DC voltage V_o from the AC input voltage v_g while operating with a nearly uniform PF. The inductor current i_L is detected by means of a current sensor, and the error signal $i_{\text{ref}} - i_L$ is sampled at the switching frequency and processed by the current controller. The outer voltage controller (dashed structure in the figure) is not included in the model.

2. System description

Fig. 1 shows a schematic circuit diagram of a PFC boost converter designed to produce a regulated DC voltage V_o from the AC voltage source v_g while operating with a nearly uniform power factor. As shown in the figure, the inductor current i_L is measured by means of a current sensor, and the error current, represented by the difference $i_{\text{ref}} - i_L$, is sampled at the switching frequency and thereafter processed in the sample-and-hold current controller. The output of this controller i_c is connected to the non-inverting input of the comperator and a suitable saw-tooth signal i_a is applied to the inverting input of the comperator. This is used to control the operation of the switch S. This operation is arranged in such a way that the switch S switches ON at the beginning of each switching period cycle and switches OFF whenever the ramp voltage i_a exceeds the control signal i_c. The diode D is activated in a manner complementary to the activation of S. The capacitor with capacitance C and the resistor with resistance R represent a smoothing filter that provides a smooth autonomous component to the dynamics of the system.

In order to regulate the output voltage v_o to a desired value V_o an outer voltage control loop is used (the structure shown dashed in Fig. 1). The crossover frequency of this controller must be of the order of some Hz for the ripple associated with the voltage feedback to be small, and the harmonic distortion of the line current acceptable.

During steady state operation, the converter is designed to operate with periodic wave forms for the various state variables. The system is subjected to two external periodicities: the switching period T and the line period $T_l \gg T$. To ensure that the average value of the inductor current is in phase with the input voltage, the reference current i_{ref} is given by the input voltage $v_g(t)$ multiplied by a suitable conductance g:

$$i_{\text{ref}}(t) = g\, v_g(t). \tag{1}$$

Here the conductance g is provided by the output of the external voltage controller. The signal

$$v_g(t) = V_g|\sin(\omega t)| \tag{2}$$

is the rectified sinusoidal source voltage with the constant amplitude V_g and the angular frequency $\omega = 2\pi/T_l$. In practice, the value of g is the conductance given by an external loop controller regulating the output voltage v_o and which is related to the rated power of the converter. However, in most applications this conductance as well as the output voltage can be considered as constant ($g = G$ and $v_o = V_o$). Indeed, in practice the output voltage is regulated to constant DC value by means of an appropriate controller. Since the small ripple in this output voltage has no effect on the phenomena reported in this study, the ripple of this voltage and its controller can be neglected. This approximation will also allow us to obtain the time varying map in closed form. Therefore, in the following, to simplify the analysis we shall ignore the outer loop. Hence, the inductor current represents the only state variable of our model.

Among many control algorithms that can be used for performing the PFC operation, in this work we choose a simple control strategy in which the value of the duty cycle $d(nT)$ is programmed digitally according the following law

$$d(nT) = d_n = \frac{1}{I_\ell}(-i_{\rm ref}(nT) + i_L(nT) + I_\ell) \tag{3}$$

where I_ℓ as defined in Fig. 1 is the amplitude of the ramp signal. It should be noted that because the duty cycle is defined cycle by cycle, its expression in (3) has a physical meaning only in the interval $(0, 1)$. If that expression gives values outside this interval, the duty cycle will be saturated to 0 or 1.

The behavior of the considered converter is described by the differential equation

$$\frac{\mathrm{d}x}{\mathrm{d}t} = \begin{cases} \dfrac{v_g(t)}{L} & \text{if } \delta(t) = 1 \\ \dfrac{v_g(t)}{L} - \dfrac{V_o}{L} & \text{if } \delta(t) = 0 \end{cases} \tag{4}$$

where $x = i_L$, while $\dfrac{v_g(t)}{L}$ and $\dfrac{V_o - v_g(t)}{L}$ are the time varying rising and falling slopes of the inductor current within a switching cycle. In Eq. (4) $\delta(t)$ is the driving binary signal with the duty cycle $d(t)$ applied to the switch which is generated by $T-$periodic PWM process, V_o is the output voltage.

3. Time varying map

Let us suppose that t is such that $\sin(\omega t) > 0$ and therefore $|\sin(\omega t)| = \sin(\omega t)$. During the ON phase ($\delta(t) = 1$), i.e., within the interval $nT < t < t_n$, $t_n = (n + d_n)T$, Eq. (4) has the form

$$\frac{\mathrm{d}x}{\mathrm{d}t} = \frac{V_g}{L}\sin(\omega t), \quad x(nT) = x_n, \tag{5}$$

with the solution

$$x(t) = x_n - \frac{V_g}{L\omega}\left[\cos(\omega t) - \cos(\omega nT)\right], \tag{6}$$

or, at the switching time $t = t_n$,

$$x(t_n) = x_n - \frac{V_g}{L\omega}\left[\cos(\omega t_n) - \cos(\omega nT)\right]. \tag{7}$$

At the time $t = t_n$, the converter switches to the phase OFF, and in the subsequent time interval $t_n \leqslant t \leqslant (n+1)T$, Eq. (4) takes the form

$$\frac{\mathrm{d}x}{\mathrm{d}t} = \frac{V_g}{L}\sin(\omega t) - \frac{V_o}{L}, \tag{8}$$

with the solution

$$x(t) = x(t_n) - \frac{V_o}{L}(t - t_n) - \frac{V_g}{L\omega}[\cos(\omega t) - \cos(\omega t_n)]. \tag{9}$$

Hence, for $t = (n+1)T$ we obtain

$$x_{n+1} = x(t_n) - \frac{V_o}{L}(1 - d_n)T - \frac{V_g}{L\omega}[\cos(\omega(n+1)T) - \cos(\omega t_n)], \tag{10}$$

with $x_{n+1} = x((n+1)T)$. Substituting the expression (7) into (10), we obtain the map for the first half line cycle, i.e., for $\omega nT \leqslant \pi$:

$$x_{n+1} = x_n - \frac{V_o}{L}(1 - d_n)T - \frac{V_g}{L\omega}\left[\cos(\omega(n+1)T) - \cos(\omega nT)\right]. \tag{11}$$

For the remaining half cycle the calculations can be done similarly by inverting the sign of the sine function. In this way we obtain

$$x_{n+1} = \begin{cases} x_n - \frac{V_o}{L} z_n T - \frac{V_g}{L\omega} \cdot [\cos(\omega(n+1)T) - \cos(\omega nT)] & \text{if } \sin(\omega nT) \geqslant 0 \\ x_n - \frac{V_o}{L} z_n T + \frac{V_g}{L\omega} \cdot [\cos(\omega(n+1)T) - \cos(\omega nT)] & \text{if } \sin(\omega nT) < 0 \end{cases} \tag{12a}$$

where

$$z_n = 1 - d_n = \begin{cases} 0 & \text{if } x_n < r_n, \\ \frac{r_n - x_n}{I_\ell} & \text{if } r_n \leqslant x_n \leqslant r_n + I_\ell, \\ 1 & \text{if } x_n > r_n + I_\ell \end{cases} \tag{12b}$$

and $r_n = i_{\text{ref}}(nT) = gV_g \sin(\omega nT)$.

Let the ratio between the line period and the switching period be $m = T_l/T$. In practice the line period is much higher than the switching period and therefore m takes high values. For instance, the set of frequency values used in this paper leads to $m = 1000$. Under these conditions, Eq. (12a) can be rewritten as

$$x_{n+1} = \begin{cases} x_n - \frac{V_o}{L} z_n T - \frac{V_g}{L\omega}\left[\cos\left(2\pi\frac{n+1}{m}\right) - \cos\left(2\pi\frac{n}{m}\right)\right] & \text{if } \sin\left(2\pi\frac{n}{m}\right) \geqslant 0 \\ x_n - \frac{V_o}{L} z_n T + \frac{V_g}{L\omega}\left[\cos\left(2\pi\frac{n+1}{m}\right) - \cos\left(2\pi\frac{n}{m}\right)\right] & \text{if } \sin\left(2\pi\frac{n}{m}\right) < 0 \end{cases} \tag{13}$$

Eq. (13) defines a non-autonomous 1D map which can easily be transformed into an autonomous 2D map by extending the state space and considering the discrete time n as a new state variable. As sine and cosine functions in Eq. (13) are m-periodic, this new variable can be considered modulo m:

$$x_{n+1} = \begin{cases} x_n - \frac{V_o}{L} z_n T - \frac{V_g}{L\omega}\left[\cos\left(2\pi\frac{k_n+1}{m}\right) - \cos\left(2\pi\frac{k_n}{m}\right)\right] & \text{if } \sin\left(2\pi\frac{k_n}{m}\right) \geqslant 0 \\ x_n - \frac{V_o}{L} z_n T + \frac{V_g}{L\omega}\left[\cos\left(2\pi\frac{k_n+1}{m}\right) - \cos\left(2\pi\frac{k_n}{m}\right)\right] & \text{if } \sin\left(2\pi\frac{k_n}{m}\right) < 0 \end{cases} \tag{14a}$$

$$k_{n+1} = (k_n + 1) \bmod m, \quad k_0 = 0 \tag{14b}$$

Moreover, when m is an even number, the map (14a) can be further simplified by restricting the variable k to the interval $\mathbb{K} = [0, m/2 - 1]$. Indeed, since $\forall\phi : \cos(\phi) = -\cos(\phi - \pi)$, the

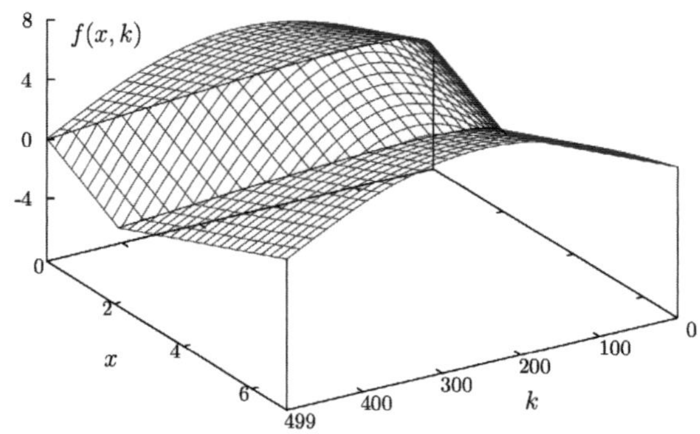

Figure 2. A typical shape of the function f. As one can see, for each fixed k the function $f(x, k^*)$ as a function of x is bimodal. One period of $f(x, k)$ with respect to k is shown. Parameters: $I_\ell = 2.1$A, $m = 1000$.

expressions for $\sin(2\pi k_n/m) \geqslant 0$ and $\sin(2\pi k_n/m) < 0$ can be unified. This leads to the final expression of our model

$$x_{n+1} = f(x_n, k_n) = x_n + \frac{V_g}{L\omega}\cos\left(2\pi\frac{k_n}{m}\right) - \frac{V_g}{L\omega}\cos\left(2\pi\frac{k_n+1}{m}\right) - \frac{V_o}{L}z_n T, \tag{15a}$$

$$k_{n+1} = h(k_n) = (k_n + 1) \bmod \frac{m}{2}, \quad k_0 = 0 \tag{15b}$$

where z_n is given by Eq. (12b) and r_n can be rewritten as follows:

$$r_n = i_{\text{ref}}(k_n) = gV_g \sin\left(2\pi\frac{k_n}{m}\right). \tag{15c}$$

Note the variable k corresponds to the discrete time restricted to the interval $\mathbb{K}$ and takes integer values

$$k_n = n \bmod \frac{m}{2}. \tag{16}$$

After extending the state space by considering the discrete time k_n as a new state variable, the model in (15) represents an autonomous triangular map where f in Eq.(15a) depends both on x and k, while h in Eq.(15b) depends on k only. Because of the possible saturation of the duty cycle (see Eq.(12b)), the phase space $\mathbb{R}_+ \times \mathbb{K}$ of map (15) consists of three partitions separated by the following two borders

$$s^- = \{(x, k) \mid x = r_k\} \tag{17a}$$

$$s^+ = \{(x, k) \mid x = r_k + I_\ell\}. \tag{17b}$$

On the outer partitions the function f is linear and increasing with respect to x, while on the middle partition it is non-linear and decreasing, so that for each fixed k, f as a function of x is bimodal, as illustrated in Fig. 2. It is worth noticing that this kind of maps results from the modeling of several other classes of power-converters (see [17] for an example).

4. Numerical simulations

Let us fix the following set of parameter values, determined by physical parameters of the considered circuit: $g = 0.01$ S, $V_g = 220\sqrt{2}$ V, $\omega = 2\pi \cdot 50$ rad/s, $V_o = 380$ V, $L = 150$ μH,

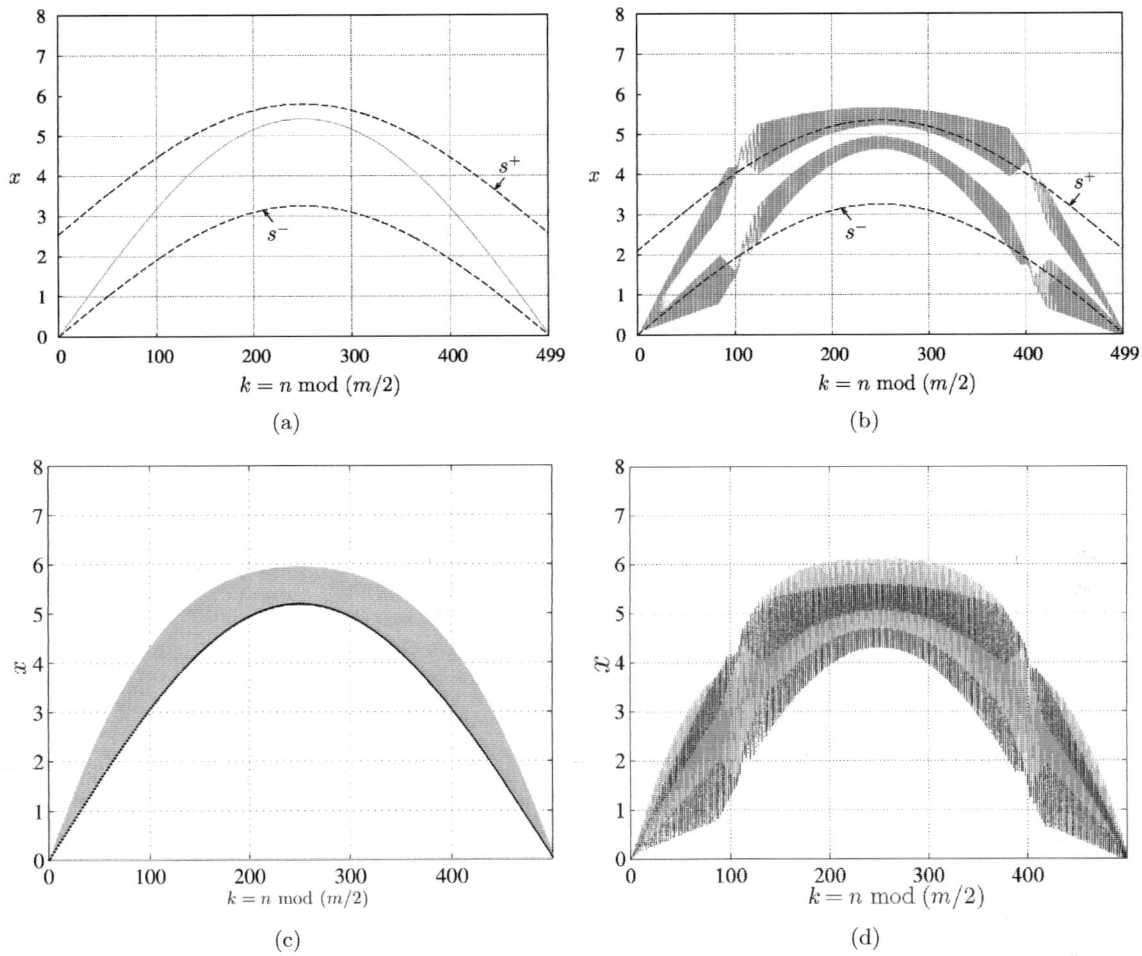

Figure 3. (a) A stable cycle and (b) a two-band chaotic attractor of map (15). Additionally, in (a) and (b) the boundaries $s^{\pm}$ are shown. In (c) and (d) the time-series obtained in continuous-time simulations at the same parameter values as in (a) and (b) are shown, dots indicate the values obtained by sampling the time series with the switching frequency. Parameters: (a), (c): $I_\ell = 2.54$A, (b), (d): $I_\ell = 2.1$A.

$T = 2\ \mu$s, and explore the possible dynamical behaviors that the system exhibits under variation of the parameter I_ℓ (recall that this parameter determines the amplitude of the ramp function). Fig. 3 shows the system behavior in the state space (k_n, x_n) for two different values of I_ℓ. In Fig. 3(a) an example for periodic dynamics of map (15) is presented, corresponding to the desired operational regime of the circuit. In this case the controller is able to impose a near unity PF for the system since, in average, the inductor current is proportional to the rectified sinusoidal input voltage without any irregular oscillation. However, this is not the case for Fig. 3(b) which shows a chaotic attractor of map (15).

For comparison, the system is also simulated using a circuit-based continuous-time switched model implemented in PSIM [18]. The resulting wave forms are shown in Fig. 3(c)-(d). As one can see, the results of continuous-time and discrete-time simulations are in a good agreement.

Journal of Physics: Conference Series **692** (2016) 012003 doi:10.1088/1742-6596/692/1/012003

5. Chaotic attractors

As inspection of the figure shows, the chaotic attractor in Fig. 4(a) is located inside the absorbing area confined by the curves defined by the images of the borders:

$$c^{0,\pm} = \left\{(x,k) \mid x = f\left(k^{(-1)}, s^{\pm}_{k^{(-1)}}\right)\right\} \tag{18a}$$

$$c^{1,\pm} = \left\{(x,k) \mid x = f\left(k^{(-1)}, f\left(k^{(-2)}, s^{\pm}_{k^{(-2)}}\right)\right)\right\} \tag{18b}$$

where

$$k^{(i)} = (k+i) \bmod m/2 \tag{18c}$$

One immediately observes that for each fixed $k = \tilde{k}$ the attractor consists of two bands, which are mapped by f on the corresponding two bands at the next value $k = \tilde{k}^{(1)}$. It is worth noticing that the size and the oscillations of the bands depends strongly on k. For example, for k close to $m/4$ the bands are broad and do not oscillate. By contrast, there are two intervals for k close to $m/8$ and $3m/8$ in which the bands suddenly decrease in size and high frequency oscillations appear.

The high frequency oscillations mentioned above lead to phenomena which may be misleading on the large scale. For example, for $k > 120$ the chaotic attractor shown in Fig. 4(b) seems to consist of four bands. However, it is clearly visible in the magnification that for each value of k the attractor has only one band, i.e., it is a one-band attractor. Remarkably, with increasing k the bands are cyclically rotating taking one of four possible positions, which leads to the large-scale structure misleadingly resembling a four-band attractor.

Numerical simulations suggest that by decreasing the parameter I_ℓ we observe a sequence of bifurcations forming a cascade which is similar to a bandcount doubling cascade well known for piecewise smooth 1D maps [19]. Recall that in continuous maps such a cascade is formed by merging bifurcations [20] and shows a sequence of chaotic attractors with $\mathcal{K} = p \cdot 2^i$ bands, $i = 0, 1, 2, \ldots$, accumulating to a parameter value at which a p-cycle changes its stability [19]. In the 2D map (15), a similar scenario can be observed under variation of I_ℓ, and also the Lyapunov exponents (see Fig. 5(d)) show the structure typical for such transitions to chaos. Note that in order to avoid difficulties related to the discrete-valued variable k, the Lyapunov exponents are calculated not for the map (15), but for the $(m/2)$th iterate $f^{m/2}$ of its first component f (see Eq. (15a)). It can easily be shown that $f^{m/2}$ represents a 1D stroboscopic map of map (15) and reflects its dynamics completely.

In addition, an unusual phenomenon can be observed in Fig. 5, namely that the bifurcations diagrams calculated at different fixed values of k appear incomplete and exhibit remarkable gaps. The location of these gaps depends on k and at the present, the observed phenomenon is far from being completely investigated. It can already be said that it results from two different mechanisms:

(i) As usual for bandcount doubling cascades in 1D maps, there are nested absorbing areas in the state space, and the number of their connected components determines the position in the bandcount doubling cascade. For example, at the parameter value corresponding to Fig. 4(a) there exists an overall absorbing area with one connected component confined by the first two images of the borders ($c^{i,\pm}$, $i = 0, 1$; as described above). Inside this absorbing area there exists a nested absorbing area with two connected components, confined by the first four images of the borders ($c^{i,\pm}$; $i = 0, 1, 2, 3$). This phenomenon is independent on k and explains the overall shape of the observed bifurcation diagrams.

(ii) In addition to the described phenomenon, there are oscillations whose appearance depends on k. As a consequence, at some values of k the bands of chaotic attractors fill the connected

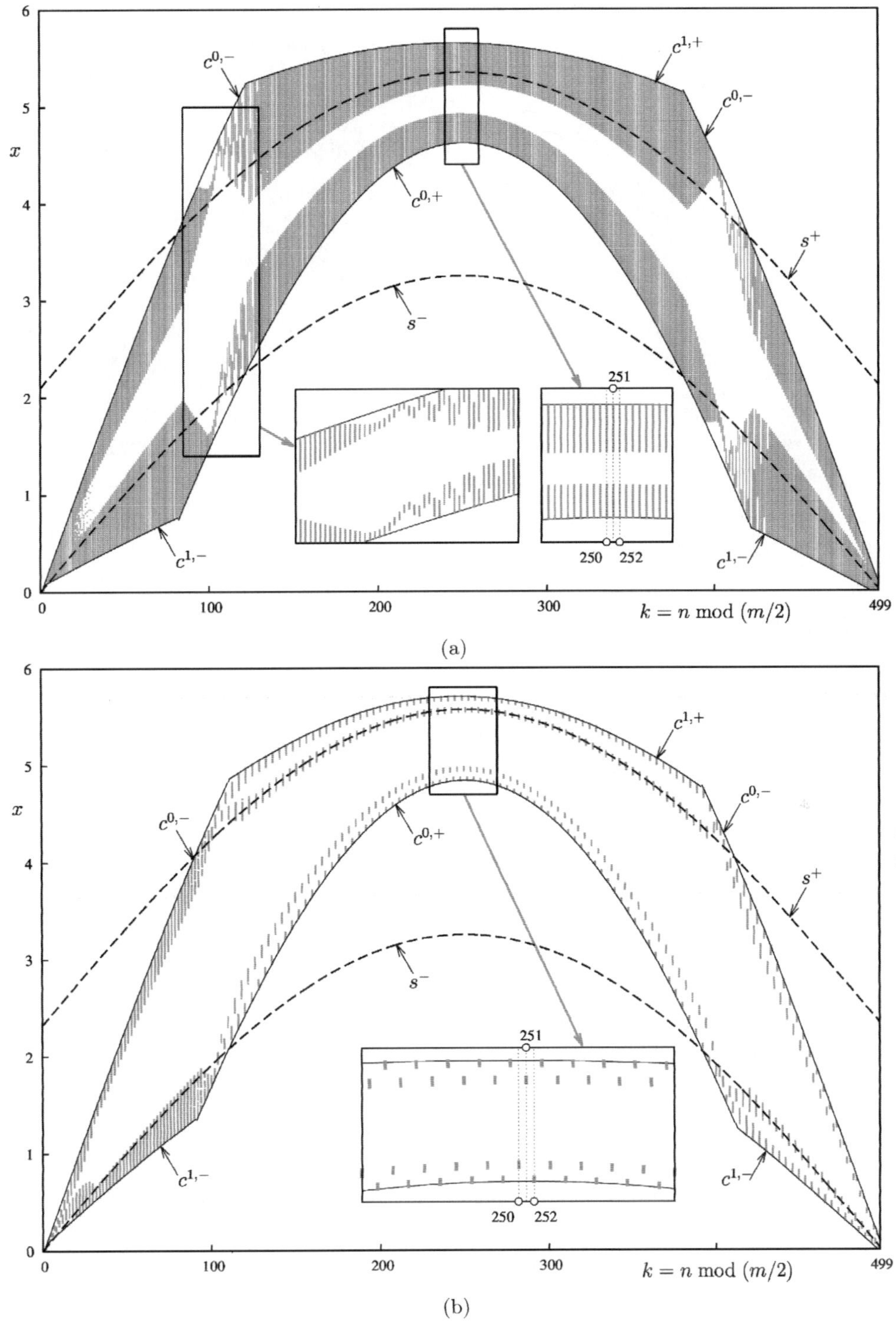

Figure 4. Chaotic attractors of map (15): (a) a two-band attractor; (b) a one-band attractor. Additionally, the borders $s^{\pm}$ and their images of rank 1 and 2 are shown. The values k = 250, 251, 252 corresponding to the bifurcation diagrams shown in Fig. 5(a), (b) and (c), respectively, are marked in the magnifications. Parameters: (a) $I_\ell = 2.1$A, (b) $I_\ell = 2.32$A. In Fig. 5 these parameter values are marked with B and C, respectively

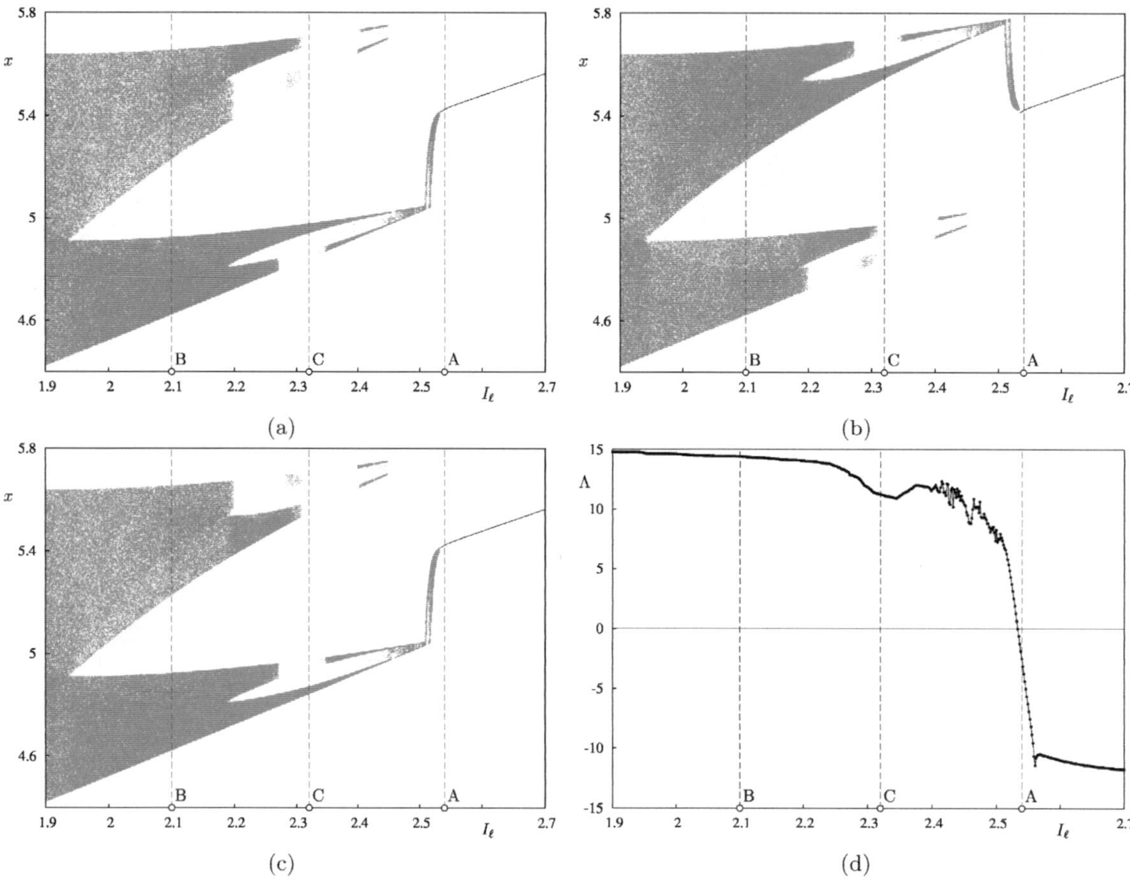

Figure 5. Bifurcation diagrams obtained under variation of I_ℓ at (a) $k = 250$, (b) $k = 251$ and (c) $k = 252$. The attractor at the parameter value marked with A is shown in Fig. 3(a), at the value marked with B in Figs. 3(b) and 4(a), and at the value marked with C in Fig. 4(b). In (d) the Lyapunov exponent Λ is shown, calculated using the $(m/2)$th iterate $f^{m/2}$.

components of the absorbing areas mentioned above, while at the others they do not. This can be observed for example in the magnifications of the attractor shown in Fig. 4(a). In the right magnification (which shows the interval $245 \leqslant k \leqslant 265$) the boundaries of the chaotic attractor are given by the first four images $c^{i,+}$; $i = 0, 1, 2, 3$ of the border. By contrast, in the left magnification (showing the interval $85 \leqslant k \leqslant 130$) the first four images of the borders are not sufficient to describe the boundaries. Similarly, for each k the band of the one-band attractor shown in Fig. 4(b) fills one of four intervals. All together, these four intervals fit to the bandcount doubling structure mentioned above. However, as at each fixed k only one of them is filled, causing the resulting bifurcation diagram calculated at this particular value of k to appear incomplete. For example, at $k = 250$ (this value is labeled in the magnification in Fig. 4(b)) the second interval from below is filled, so that only this band is present in the bifurcation diagram calculated at this value of k (see Fig. 5(a) and the parameter value labeled by the letter C). For the next values of k, i.e., for $k = 251$ and $k = 252$, the third and the first intervals from below are filled, respectively. This explains the differences between Figs. 5(a), 5(b) and 5(c).

The mechanism described above explains the appearance of gaps on the bifurcation diagrams shown Fig. 5 and shows that they are not related to numerical inaccuracies. As a next step, remaining for a future work, it is necessary to investigate the bifurcations leading the number of bands to change and to find the regularities in their occurrence.

6. Conclusions

Over the last couple of decades, active PFC circuits have become common in power supply systems in a broad range of applications such as telecommunications, computing equipment and micro-grids among others. The advantage of using active circuits is the higher efficiency and the better power quality than in passive circuits. The penalty is the added complexity that results from using switching devices working in nonlinear regime. This may cause the basic periodic operating modes to disappear through nonsmooth bifurcation phenomena, and may also introduce new oscillatory regimes in the system, leading to poor performances in terms of the power quality. In this paper an investigation of the dynamics of an AC-DC boost converter working as a PFC circuit was performed by using a nonlinear piecewise smooth time varying map. The results obtained from this map were validated by a circuit-based switched model. Nonsmooth bubbling phenomena have been observed. Numerical simulations show that when some parameter values change a sequence of bifurcations forming a cascade which is similar to a bandcount doubling cascade is observed. However, in the considered case the bifurcation scenario is more complicated, as it results from a mixture of time-independent appearance and disappearance of nested absorbing areas on one side and time-dependent bubbling-related phenomena. To our knowledge similar effects have never been reported before.

Acknowledgement

The work of V. Avrutin was partially supported by the German Research Foundation within the scope of the project *"Organizing centers in discontinuous dynamical systems: bifurcations of higher codimension in theory and applications"*.

References

[1] O. Garcia, J. A. Cobos, R. Prieto, P. Alou, J. Uceda, "Single phase power factor correction: a survey", *IEEE Trans. on Power Electronics,* vol. 18, no. 3, pp. 749-755, 2003.

[2] J. Sebastian, D. G. Lamar, M. M. Hernando, A. Rodriguez-Alonso, and A. Fernandez, "Steady-state analysis and modeling of power factor correctors with appreciable voltage ripple in the output-voltage feedback loop to achieve fast transient response", *IEEE Trans. on Power Electronics,* vol. 24, no. 11, pp. 2555-2566, 2009.

[3] H. H. C. Iu, Y. Zhou, and C. K. Tse, "Fast-scale instability in a boost PFC converter under average current control", *Int. J. of Circuit Theory and Applications,* vol. 31, pp. 611-624, 2003.

[4] M. Orabi and T. Ninomiya, "Nonlinear dynamics of power-factor-correction converter", *IEEE Trans. on Power Electronics,* vol. 50, no. 6, pp. 1116-1125, 2009.

[5] S. C. Wong, C. K. Tse, M. Orabi and T. Ninomiya, "The method of double averaging: an approach for modeling powerfactor-correction switching converters", *IEEE Trans. on Circuits and Systems I,* vol. 53, no. 2, pp. 454-462, 2006.

[6] W. Lu, Sh. Lang, L. Zhou, H. H.-Ch. Iu and T. Fernando, "Improvement of Stability and Power Factor in PCM Controlled Boost PFC Converter With Hybrid Dynamic Compensation", *IEEE Trans. on Circuits and Systems I,* vol. 62, no. 1, pp. 320-328, 2015.

[7] W. Cheng, J. Song, H. Li and Y. Guo, "Time-Varying Compensation for Peak Current-Controlled PFC Boost Converter", *IEEE Trans. on Power Electronics,* vol. 30, no. 6, pp. 3431-3437, 2015.

[8] A. El Aroudi, M. Orabi and L. Martinez-Salamero, "A representative discrete-time model for uncovering slow and fast scale instabilities in boost power factor correction AC-DC pre-regulators", *Int. J. of Bifurcation and Chaos,* vol. 18, no. 10, pp. 3073-3092, 2008.

[9] G. Chu, C. K. Tse, and S-C Wong, "Line-frequency instability of PFC power supplies", *IEEE Trans. on Power Electronics,* vol. 24, no. 2, pp. 469-482, 2009.

[10] A. El Aroudi and M. Orabi, "Stabilizing technique for AC/DC boost PFC converter based on time delay feedback",*IEEE Trans. on Circuits and Systems II: Express Briefs,* vol. 57, no. 1, pp. 56-60, 2010.

[11] A. El Aroudi and M. Orabi, "Dynamics of PFC power converters subject to time-delayed feedback control", *Int. J. of Circuit Theory and Applications,* vol. 40, no. 1, pp. 15-35, 2012.
[12] J. H. B. Deane and D. C. Hamill, "Instability, subharmonics, and chaos in power electronic systems", *IEEE Trans. on Power Electronics,* vol. 5, no. 2, pp. 260-268, Jul. 1990.
[13] D. C. Hamill, J. H. B. Deane, and D. Jefferies, "Modeling of chaotic DC-DC converters by iterated nonlinear mappings", *IEEE Trans. on Power Electronics,* vol. 7, no. 1, pp. 25-36, 1992.
[14] V. S. Baushev and Z. T. Zhusubaliyev, "Indeterminable states of a voltage regulator with pulse-width control", *Electrical Technology* vol.3, pp. 85-88, 1992.
[15] B. Robert and C. Robert, "Border Collision Bifurcations in a one-dimensional piecewise smooth map for a PWM current-programmed H-bridge inverter", *Int. J. of Control,* vol. 7, no. 16, pp. 1356-1367, 2002.
[16] Z. T. Zhusubaliyev, E.Mosekilde, A. I Andriyanov and V. V. Shein V.V, "Phase synchronized quasiperiodicity in power electronic inverter systems", *Physica D: Nonlinear Phenomena,* vol. 268, pp. 14-24, 2014.
[17] V. Avrutin, E.Mosekilde, Z.T. Zhusubaliyev and L. Gardini, "Onset of chaos in a single-phase power electronic inverter", *Chaos: An Interdisciplinary J. of Nonlinear Science,* vol. 25, pp. 043114, 2015.
[18] `http://www.powersimtech.com/`
[19] V. Avrutin and I. Sushko, "A gallery of bifurcation scenarios in piecewise smooth 1D maps", *Global Analysis of Dynamic Models in Economics, Finance and the Social Sciences,* Springer, pp. 369-395, 2015.
[20] V. Avrutin, L. Gardini, M. Schanz and I. Sushko, "Bifurcations of chaotic attractors in one-dimensional maps", *Int. J. of Bifurcation and Chaos,* vol. 24, no. 8, pp. 1440012, 2014.

Homomorphism to $\mathbb{R}$ generated by abstract length functions: a dynamical construction

Gastão Bettencourt[1] **and Sérgio Mendes**[2]

[1]Departamento de Matemática, Fac. Ciências, Universidade da Beira Interior, Covilhã, Portugal, [2]Departamento de Matemática, ISCTE - Lisbon University Institute, Lisbon, Portugal

E-mail: gastao@ubi.pt, sergio.mendes@iscte.pt

Abstract. Erschler and Karlsson in [5] construct a homomorphism of a finitely generated group G to $\mathbb{R}$ using a random walk approach. Central to their construction were the word length ℓ and a well behaved measure μ on G. In this note we define a class of abstract length functions and prove that Erschler and Karlsson construction can also be applied to this class.

1. Introduction

It was Pólya [11], in 1921, who started the study of random walks on $\mathbb{Z}^n$. It took almost 40 years, until Kesten [10], in 1959, started the study of random walks on infinite, finitely generated groups. After that, in 1972 Avez [1] introduced the entropy h of a random walk on a finitely generated group, arguably the most important numerical invariant associated to a group endowed with a probability measure, and connected entropy with the growth rate v of the group. In his subsequent work [2], Avez put in evidence the strong connection between entropy and the existence of bounded harmonic functions on the group. The drift ℓ is another number that has emerged in the study of random walks. These numbers are linked via the fundamental inequality popularized by Vershik [12], $h \leqslant v\ell$. For a comprehensive reference in a more general setting see Kaimanovich and Vershik [9].

The length function of a finitely generated group is a well established concept in geometric group theory. An abstract length function f, for a group G, associates to an element $g \in G$ a real number subject to certain axioms. In an attempt to understand cancelation properties for elements of free groups, Lyndon [7] defined integer valued length functions. Later, Harrison [6] used Lyndon axioms to define real abstract length functions on groups.

In this note we consider abstract length functions which are subject to the only basic axioms of being subadditive and inverse invariant. Hence, we consider a broader class of length functions which includes Lyndon's length functions. Our class of abstract length functions give rise to a pseudometric in G, that is, we admit some nonidentity elements to have (abstract) length zero. The set of the elements $\{g \in G : f(g) = 0\}$ actually has the structure of a subgroup of G which we conveniently call the kernel of f. Some of its properties are studied in section 3.

There are several ways to construct nontrivial homomorphisms, depending on the setting. For example, if G is amenable any homogeneous quasimorphism to $\mathbb{R}$ is in fact a homomorphism, see [4]. Karlsson and Ledrappier in [8] proved that if a finite first moment random walk has zero entropy and positive drift, then there exists a non-trivial homomorphism to $\mathbb{R}$. Erschler and

Journal of Physics: Conference Series **692** (2016) 012004 doi:10.1088/1742-6596/692/1/012004

Karlsson [5] used a dynamical construction to establish the existence of such homomorphism in the case where the drift can also be zero. In [3], the authors followed Erschler and Karlsson construction to define a homomorphism to $\mathbb{R}$ for a certain class of semidirect products. The original construction starts by making a judicious choice of a sequence of functions T_n, and of a numerical sequence $\alpha(n)$ such that the product $\alpha(n)T_n$ will have a sublimit T which is precisely the desired homomorphism.

We give now a brief description of this note. In section 2 we define abstract length functions on a group G and give some examples. In section 3, we collect some elementary properties of abstract length functions. Finally, in section 4 we focus on a dynamical way of obtaining homomorphisms from a finitely generated group G to $\mathbb{R}$. Specifically, we show that Erschler and Karlsson construction still works for the broader class of abstract length functions. An interesting question relating this construction with the notion of the kernel of a subadditive inverse invariant functions f (see sections §2, 3) is to find under what conditions the kernel K_f equals the kernel of the homomorphism $T = \lim_k T^{\alpha}_{n_k}$, see Remark (4.7).

The authors would like to thank the anonymous referees for their suggestions which helped to improve the paper.

2. A class of subadditive functions

Let $f : G \to \mathbb{R}$ be a real function on a group G which is subadditive and inverse invariant, i.e.

$$f(gh) \leqslant f(g) + f(h) \tag{1}$$

$$f(g^{-1}) = f(g). \tag{2}$$

These properties are a subset of the properties that characterize the usual length function on groups, $\ell(g)$. We denote the set of functions verifying (1) and (2) simply by SII (subadditive inverse invariant).

Denoting by e the identity of G, we have

$$f(e) = f(gg^{-1}) \leqslant f(g) + f(g^{-1}) = 2f(g)$$

Taking $g = e$ in the above inequality we obtain $f(e) \geqslant 0$. Hence,

$$f(g) \geqslant 0$$

for every $g \in G$.

Now, suppose there is some $g_o \in G$ such that $f(g_o) = 0$. Define the following subset of G

$$K = \{g \in G : f(g) = 0\}. \tag{3}$$

Since $0 \leqslant f(e) \leqslant 2f(g_o)$ then $f(e) = 0$ and $e \in K$. Also if $g \in K$ then $g^{-1} \in K$ by (2). On the other hand, if $g, h \in K$ it follows that

$$0 \leqslant f(gh) \leqslant f(g) + f(h) = 0$$

and $gh \in K$. So we conclude the following

Proposition 2.1. *Let $f \in \mathit{SII}$ and suppose K is not empty. Then K is a subgroup of G.*

This result motivates the following definition.

Definition 2.2. *Let $f \in \mathit{SII}$. $K = \{g \in G : f(g) = 0\}$ will be called the kernel of f.*

We introduce some simple examples and remarks to elucidate these concepts.

Example 2.3. *Let $G = (\mathbb{R}^+, \cdot)$ be the multiplicative group of positive real numbers and let $f : G \to \mathbb{R}$ be the characteristic function of the set of positive irrational numbers*

$$f(x) = \begin{cases} 0 & x \in \mathbb{Q}^+ \\ 1 & x \in \mathbb{R}^+ \backslash \mathbb{Q}^+ \end{cases}$$

Since $x \in \mathbb{Q}^+$ if and only if $x^{-1} \in \mathbb{Q}^+$ it follows that f is inverse invariant. It is also easy to check subadditivity. We have $K = \mathbb{Q}^+$.

Example 2.4. *More general, let $G = (\mathbb{R}^+, \cdot)$ and let $f : G \to \mathbb{R}$ be defined by*

$$f(x) = \begin{cases} \alpha & x \in \mathbb{Q}^+ \\ \beta & x \in \mathbb{R}^+ \backslash \mathbb{Q}^+ \end{cases}$$

If $0 \leqslant \alpha \leqslant 2\beta$ then $f \in SII$, which generalizes last example. Regarding the kernel, we have
- *$\beta = 0$ then $f = 0$ and $K = \mathbb{R}^+$*
- *$\alpha = 0$ and $\beta > 0$ then $K = \mathbb{Q}^+$*
- *$\alpha > 0$ then $K = \emptyset$.*

Here is a nonabelian example.

Example 2.5. *Let $G = \left\{ \begin{pmatrix} 1 & x & y \\ 0 & 1 & z \\ 0 & 0 & 1 \end{pmatrix} : x, y, z \in \mathbb{Z} \right\}$ denote the discrete Heisenberg group and let $f : G \to \mathbb{R}$ be defined by*

$$f(g) = |x + z|.$$

We see that $f \in SII$ and the kernel is

$$K = \left\{ \begin{pmatrix} 1 & x & y \\ 0 & 1 & -x \\ 0 & 0 & 1 \end{pmatrix} : x, y \in \mathbb{Z} \right\}$$

We note that K is a normal subgroup of G.

Remark 2.6. *It is interesting to note that if we consider a subadditive function which instead of (2) verifies $f(g^{-1}) = f(g)$ only for $g \in K$ we still have that K is a subgroup. Other properties still remain valid, e.g. see (3.1), (3.2).*

Some ways to generate new inverse invariant subadditive functions are given by the next three results, whose proof is straightforward.

Proposition 2.7. *Let $f_1, f_2 \in SII$, $\alpha > 0$ and $\varphi : H \to G$ be a group homomorphism. We have*

- *$\alpha f_1 \in SII$*
- *$f_1 + f_2 \in SII$*
- *$f_1 \circ \varphi \in SII$*

Proposition 2.8. *Let $f \in SII$ and $\varphi : G \to G$ a group homomorphism. Then we have*

$$\frac{1}{n} \sum_{k=0}^{n-1} f \circ \varphi^k \in SII$$

Proposition 2.9. *Let $f_k \in SII$ for all $k \in \mathbb{N}$ and $\varphi : G \to G$ a group homomorphism. We have*

$$\frac{1}{n} \sum_{k=0}^{n-1} f_k \circ \varphi \in SII$$

Journal of Physics: Conference Series **692** (2016) 012004 doi:10.1088/1742-6596/692/1/012004

We collect now some random remarks. The first connects elements of SII with the familiar concept of distance, the second puts in evidence that subadditivity is the fundamental concept in the defining equations (1) and (2), the third remark justify why the class SII is somehow unusual to a reader familiar with group theory and homomorphisms, and finally, the last one connects simple groups, a subclass of functions of SII and bi-invariant metrics. The first time we saw this result was in the work of Polterovich.

Remark 2.10. *If it happens that $f(e) = 0$, a left invariant pseudo metric d on G is defined as $d(g, h) = f(g^{-1}h)$. If the kernel K of f is nonempty and trivial, then f defines a metric.*

Remark 2.11. *We also note that if we have a subadditive function f that is not inverse invariant we can always construct an inverse invariant subadditive function by making*

$$\widetilde{f}(g) = \frac{f(g) + f(g^{-1})}{2}.$$

Remark 2.12. *In the context of homomorphisms the concept of inverse invariance is void, that is, if $\psi : G \to \mathbb{R}$ is a homomorphism to $\mathbb{R}$ which is inverse invariant, then ψ is trivial. In fact,*

$$0 = \psi(e) = \psi(gg^{-1}) = 2\psi(g)$$

so, $\psi(g) = 0$ for every $g \in G$.

Remark 2.13. *Assume now that the group G is simple and suppose that $f \in SII$ is also conjugacy invariant, in the sense $f(sgs^{-1}) = f(g)$, for all $s, g \in G$. In particular for $h \in K$, $g \in G$ we have $f(ghg^{-1}) = f(h) = 0$, and $ghg^{-1} \in K$, that is, K is normal on G. This means that either $K = G$, which amounts to say that f is trivial, or $K = \{e\}$. Regarding the first remark, keeping in mind that a right invariant metric is generated by a conjugacy invariant function of SII, we can say that a bi-invariant non trivial pseudo metric on a simple group is indeed a metric.*

3. Some properties of the kernel

We recall that when $\phi : G \to H$ is a group homomorphism, and h is an element of the kernel then, for every $g \in G$ we have

$$\phi(gh) = \phi(g)\phi(h) = \phi(g).$$

In our setting we still have the same result. For the remaining part of this section, G is an arbitrary group and $f \in SII$. Suppose that K is nonempty and let $h \in K$. We have, for every $g \in G$,

$$f(g) = f(ghh^{-1}) \leqslant f(gh) + f(h^{-1}) \leqslant f(gh).$$

Since

$$f(gh) \leqslant f(g) + f(h) = f(g)$$

it follows that $f(gh) = f(g)$.

Conversely, if $f(gh) = f(g)$, for every g, then

$$f(eh) = f(e)$$

and since K is nonempty, $f(e) = 0$, and $h \in K$. We have established the following criteria for the elements of K

Proposition 3.1. *Let K be nonempty. $h \in K$ if, and only if, $f(gh) = f(g)$ for every $g \in G$.*

Evidently we also have the same statement regarding $f(hg) = f(g)$. A condition of normality of K is available:

Journal of Physics: Conference Series **692** (2016) 012004 doi:10.1088/1742-6596/692/1/012004

Proposition 3.2. *Let K be nonempty. K is normal on G if, and only if, for every $g, \overline{g} \in G$, $h \in K$ we have $f(gh\overline{g}) = f(g\overline{g})$*

Proof. If K is normal on G we have $ghg^{-1} \in K$ for $g \in G, h \in K$. So $gh = h'g$ and $f(gh\overline{g}) = f(h'g\overline{g}) = f(g\overline{g})$ where in the last identity we used proposition (3.1). On the other hand, if the condition is fulfilled then for $\overline{g} = g^{-1}$, we have $f(ghg^{-1}) = f(gg^{-1}) = f(e)$. But $f(e) = 0$ since K is nonempty. In this way we see that $ghg^{-1} \in K$ and K is normal on G. □

Now suppose we have two elements g, h such that $f(gh) = 0$. What can be said about $f(g)$ or $f(h)$? To answer this question, just notice that

$$f(g) = f(ghh^{-1}) \leqslant f(gh) + f(h) = f(h)$$

$$f(h) = f(g^{-1}gh) \leqslant f(g) + f(gh) = f(g)$$

Hence, we have the following

Proposition 3.3. *If $gh \in K$ then $f(g) = f(h)$.*

This result is obvious for $g, h \in K$. The interesting point is that it works for all $g, h \in G$, and in particular for all $g, h \in G \backslash K$. For concreteness let $a, b \in G$ such that $ab \in K$ and neither a and b belongs to K. Let $A = \{a, b, a^{-1}, b^{-1}\}$. The previous result can be reformulated as

Proposition 3.4. *Let $a, b \in G \backslash K$ such that $ab \in K$, and A as above. We have $f(A) = \{f(a)\}$*

In the same vein of reasoning we can ask what are the possible images of $f(xy)$ for $x, y \in A$. We have

Proposition 3.5. *Let $a, b \in G \backslash K$ such that $ab \in K$, and A as above. We have*

$$f(A^2) \subset \{0, f(a^2), f(b^2), f(ba)\}.$$

Proof. Enumerating all possible cases we have
Case 1, $(x = a)$:

$$\{f(ay) : y \in A\} = \{f(aa), f(aa^{-1}), f(ab), f(ab^{-1})\} = \{f(a^2), f(e), 0, f(ab^{-1})\}$$

Using proposition (3.1)

$$f(ab^{-1}) = f(abb^{-1}b^{-1}) = f(b^{-2}) = f(b^2)$$

and we have

$$\{f(ay) : y \in A\} = \{0, f(a^2), f(b^2)\}$$

Case 2, $(x = a^{-1})$:

$$\{f(a^{-1}y) : y \in A\} = \{f(e), f(a^{-1}a^{-1}), f(a^{-1}b), f(a^{-1}b^{-1})\}$$

Again, by proposition (3.1)

$$f(a^{-1}b) = f(a^{-1}a^{-1}ab) = f(a^2) \text{ and } f(a^{-1}b^{-1}) = f(ba).$$

In this case we have

$$\{f(a^{-1}y) : y \in A\} = \{0, f(a^2), f(ba)\}$$

Case 3, $(x = b)$:

$$\{f(by) : y \in A\} = \{f(ba), f(ba^{-1}), f(b^2), 0\} = \{f(ba), f(ab^{-1}), f(b^2), 0\} = \{f(ba), f(b^2), 0\}$$

Journal of Physics: Conference Series **692** (2016) 012004 doi:10.1088/1742-6596/692/1/012004

Case 4, ($x = b^{-1}$):

$$\{f(b^{-1}y) : y \in A\} = \{f(b^{-1}a), f(b^{-1}a^{-1}), 0, f(b^{-2})\}.$$

Since $f(b^{-1}a) = f(a^{-1}b) = f(a^2)$, we have

$$\{f(b^{-1}y) : y \in A\} = \{f(a^2), f(ab), f(b^2), 0\}.$$

Summing up we have the result. □

If a and b are as in the former proposition but in addition, $f(ba) = 0$, the values that $f(xy)$ can attain are even more restricted. In this case we have

$$f(b^2) = f(bb) = f(ba^{-1}ab) = f(ba^{-1}) = f(baa^{-1}a^{-1}) = f(a^2).$$

This is resumed as

Corollary 3.6. *Let $a, b \in G\backslash K$ such that $ab \in K$, and A as above. Assume $ba \in K$. We have*

$$f(A^2) \subset \{0, f(a^2)\}.$$

We note that if we have $a, b \in G\backslash K$ such that $ab \in K$, and if a and b commute we have $f(ab) = f(ba)$. Another situation where this can also happen is when K is a normal subgroup of G. Just note that $ab \in K \lhd G$ means that for all $g \in G$ $gabg^{-1} = h' \in K$. Let $g = a^{-1}$. We have $f(a^{-1}aba) = f(h') = 0$, that is $f(ba) = 0$.

Corollary 3.7. *Let $a, b \in G\backslash K$ such that $ab \in K$, A as above and $K \lhd G$. We have*

$$f(A^2) \subset \{0, f(a^2)\}$$

These type of results are generalizable. One typical result is given by the next reasoning. Assume that $a, b, c \in G\backslash K$ are such that $f(abc) = 0$. We have

$$f(a) = f(abcc^{-1}b^{-1}) = f(c^{-1}b^{-1}) \leqslant f(b) + f(c)$$

that is

$$f(a) - f(c) \leqslant f(b).$$

We also have

$$f(c) = f(b^{-1}a^{-1}abc) = f(b^{-1}a^{-1}) \leqslant f(a) + f(b),$$

which implies

$$f(c) - f(a) \leqslant f(b).$$

Hence, we have proved the following:

Proposition 3.8. *Let $a, b, c \in G\backslash K$ such that $abc \in K$. We have*

$$|f(a) - f(c)| \leqslant f(b).$$

4. A dynamical construction

From now on we restrict to infinite, finitely generated groups G, endowed with a probability measure μ. Denote by e the identity of G. Let $S = \{s_1, \cdots, s_p\} \subset G$ be a finite set of generators. The word length of g is defined to be

$$l(g) = \min\{n \in \mathbb{N} : g = s_1 \cdots s_n,\ s_i \in S\}.$$

The set of generators is called symmetric if $X^{-1} = X$. In that case the word length is generated in the usual way by the distance associated with the Cayley graph of G and satisfy the ususal properties:

$$l(gh) \leqslant l(g) + l(h). \tag{4}$$

$$l(g^{-1}) = l(g). \tag{5}$$

Definition 4.1. *We say that μ has finite first moment if*

$$\sum_{g \in G} l(g)\mu(g) < +\infty$$

Given probability measures μ and ν on G, define the convolution measure as

$$\mu * \nu(g) = \sum_{h \in G} \mu(h) h\nu(g)$$

$$= \sum_{h \in G} \mu(h)\nu(h^{-1}g)$$

where $h\nu(g) = \nu(h^{-1}g)$. Denote by μ^{*n} the n-fold convolution.

Using the word length define a sequence L_n as

$$L_n = \sum_{h \in G} l(h)\mu^{*n}(h). \tag{6}$$

The sequence L_n is subadditive, $L_{n+m} \leq L_n + L_m$. Hence, the classical Fekete lemma ensures that the limit

$$\ell_\mu = \lim_{n \to +\infty} \frac{L_n}{n}$$

exists in $\mathbb{R} \bigcup \{-\infty, +\infty\}$. Since $l \geq 0$ and L_n is subadditive we have $L_n \leq nL_1$ and since μ has finite first moment it follows that $\ell_\mu < +\infty$.

It will be usefull to define the following map

$$K_g^n(h) = g\mu^{*n}(h) - \mu^{*n}(h) \tag{7}$$

which play the role of a kernel and verifies

$$\sum_{g \in G} K_g^n(h)\mu(g) = \mu^{*(n+1)}(h) - \mu^{*n}(h). \tag{8}$$

Now, given an abstract length function f, we may still define in analogy with (6)

$$F_n = \sum_{h \in G} f(h)\mu^{*n}(h). \tag{9}$$

We note that F_n verifies $F_{n+m} \leq F_n + F_m$, hence we have $F_n \leq nF_1$. In view of this, for an abstract length function f in G, we consider a measure μ such that

$$\sum_{g\in G} f(g)\mu(g) < +\infty$$

Therefore, we have $F_1 < +\infty$.

Following the construction of Erschler-Karlsson [5], we will define the function $T_n(g)$ as follows:

$$\begin{aligned} T_n(g) &= \sum_{h\in G}(f(gh) - f(h))\mu^{*n}(h) \\ &= \sum_{h\in G} f(h)K_g^n(h). \end{aligned} \tag{10}$$

The function T_n verifies the following properties

Lemma 4.2. *Let $T_n(g)$ be defined as above. Then*

(i) $T_n(e) = 0$

(ii) $|T_n(g)| \leqslant f(g)$

(iii) $\sum_{g\in G} T_n(g)\mu(g) = F_{n+1} - F_n$.

Proof. The proof is a simple calculation. □

$T_n(g)$ is not a homomorphism, but it will be fundamental to understand how far it is from being one. The next result bounds this difference. For the sake of readability we will sometimes use the following notation

$$\mathrm{Def}_{T_n}(g_1, g_2) = T_n(g_1g_2) - T_n(g_1) - T_n(g_2). \tag{11}$$

Lemma 4.3. *We have*

$$|\mathrm{Def}_{T_n}(g_1, g_2)| \leqslant f(g_1)\sum_{h\in G}|K_{g_2}^n(h)|. \tag{12}$$

Note that, since

$$\sum_{h\in G}|K_g^n(h)| \leq 2$$

we can consider a sequence $\beta(n)$ such that for every $g \in G$ there exists $C(g)$ with

$$\sum_{h\in G}|K_g^n(h)| \leq C(g)\beta(n).$$

In fact, the above constant $C(g)$ can be made more explicit.

Lemma 4.4. *If $\sum_{h\in G}|K_g^n(h)| \leqslant C(g)\beta(n)$ then there exists a constant C_o such that*

$$\sum_{h\in G}|K_g^n(h)| \leqslant l(g)C_o\beta(n).$$

See [3] for the proof of lemmas (4.3) and (4.4). It is convenient to introduce the following notation:

$$\gamma(n) = \max_{s\in S}|T_n(s)|.$$

We have the following

Journal of Physics: Conference Series **692** (2016) 012004 doi:10.1088/1742-6596/692/1/012004

Proposition 4.5. *Let $T_n(g)$ be defined as above, μ a probability measure on G and $g \in \langle supp\,\mu \rangle$. We have*

$$|T_n(g)| \leqslant C_1\beta(n)l^2(g) + l(g)\gamma(n). \tag{13}$$

Proof. Let $g \in \langle supp\,\mu \rangle \subset G$. Since G is generated by the finite set $S = \{s_1, \cdots, s_p\}$ we can write $g = s_1 \cdots s_m$ where $m = l(g)$. By the previous lemmas we have

$$|\mathrm{Def}_{T_n}(s_1 \cdots s_{m-1}, s_m)| \leqslant f(s_1 \cdots s_{m-1}) \sum_{h \in G} |K^n_{s_m}(h)|$$

$$\leqslant \left(\sum_{i=1}^{m-1} f(s_i) \right) \sum_{h \in G} |K^n_{s_m}(h)| \leqslant \left(\sum_{i=1}^{m-1} f(s_i) \right) C_o\beta(n).$$

Writing $f_o = \max_{s \in S} f(s)$, we get

$$|\mathrm{Def}_{T_n\,n}(s_1 \cdots s_{m-1}, s_m)| \leqslant (m-1) f_o C_o \beta(n)$$

Therefore,

$$\begin{aligned}
|T_n(g)| &= |T_n(s_1 \cdots s_m)| \\
&= |T_n(s_1 \cdots s_m) - T_n(s_1 \cdots s_{m-1}) - T_n(s_m) + \\
&\quad + T_n(s_1 \cdots s_{m-1}) - T_n(s_1 \cdots s_{m-2}) - T_n(s_{m-1}) + \cdots \\
&\quad \cdots + T_n(s_1 s_2) - T_n(s_1) - T_n(s_2) + \sum_{i=1}^{m} T_n(s_i)| \\
&\leqslant f_o C_o \beta(n)((m-1) + (m-2) + \cdots + 1) + \sum_{i=1}^{m} |T_n(s_i)| \\
&\leqslant f_o C_o \beta(n)((m-1) + (m-2) + \cdots + 1) + m\gamma(n) \\
&\leqslant f_o C_o \beta(n) m^2 + m\gamma(n).
\end{aligned}$$

□

Denote $\Delta_n L = L_{n+1} - L_n$ and $\Delta_n F = F_{n+1} - F_n$. We may now state and prove the main result of this section.

Theorem 4.6. *Suppose that μ is non-degenerate, has finite second moment and for some sequence n_k it holds that $\Delta_{n_k} F \geq \Delta_{n_k} L > 0$ and*

$$\lim_{k \to +\infty} \frac{\beta(n_k)}{\Delta_{n_k} L} = 0.$$

Then G admits a non-trivial homomorphism to $\mathbb{R}$.

Proof. By proposition (4.5) we have $|T_n(g)| \leqslant C_1\beta(n)l^2(g) + l(g)\gamma(n)$ so

$$\begin{aligned}
\sum_{g \in G} |T_n(g)|\mu(g) &\leqslant C_1\beta(n) \sum_{g \in G} l^2(g)\mu(g) + \gamma(n) \sum_{g \in G} l(g)\mu(g) \\
&= C_2\beta(n) + \gamma(n)C_3.
\end{aligned}$$

Using the hypotheses of this proposition and lemma (4.2) we have

$$0 < \Delta_{n_k} L \leqslant \Delta_{n_k} F = \sum_{g \in G} T_{n_k}(g)\mu(g) = |\sum_{g \in G} T_{n_k}(g)\mu(g)|$$

$$\leqslant \sum_{g\in G} |T_{n_k}(g)|\mu(g) \leqslant C_2\beta(n_k) + \gamma(n_k)C_3$$

Again using the hypotheses we have that for any $\varepsilon > 0$ there exists n_k such that $\beta(n_k) \leqslant \varepsilon\,\Delta_{n_k}L$ and we have

$$0 < \Delta_{n_k}L \leqslant C_2\varepsilon\,\Delta_{n_k}L + \gamma(n_k)C_3.$$

Choosing ε such that $C_2\varepsilon < 1/2$ we know that for n_k big enough we have

$$0 < (1/2)\Delta_{n_k}L \leqslant (1 - C_2\varepsilon)\Delta_{n_k}L \leqslant \gamma(n_k)C_3. \tag{14}$$

From this inequality we can draw two conclusions. First we see that for n_k sufficiently big we have

$$0 < \gamma(n_k)$$

which allows us to define the coefficient

$$\alpha(n_k) = \frac{1}{\gamma(n_k)}$$

and the second conclusion is

$$0 < \alpha(n_k)\,\Delta_{n_k}L \leqslant 2C_3.$$

So we see that

$$0 \leqslant \alpha(n_k)\beta(n_k) = \alpha(n_k)\,\Delta_{n_k}L\,\frac{\beta(n_k)}{\Delta_{n_k}L} \leqslant 2C_3\frac{\beta(n_k)}{\Delta_{n_k}L}$$

and using our convergence hypothesis we conclude that

$$\alpha(n_k)\beta(n_k) \to 0$$

Let us define

$$T_n^\alpha(g) = \alpha(n)T_n(g), \quad \text{and} \quad T(g) = \lim_{k\to+\infty} T_{n_k}^\alpha(g).$$

Using lemma (4.3) we can say that

$$|\mathrm{Def}_{T_{n_k}^\alpha}(g_1, g_2)| \leqslant f(g_1)C(g_2)\beta(n_k)\alpha(n_k)$$

and

$$\lim_{k\to+\infty} \mathrm{Def}_{T_{n_k}^\alpha}(g_1, g_2) = 0$$

which amounts to say that

$$T(g_1g_2) = T(g_1) + T(g_2).$$

□

Remark 4.7. *Let $h \in K_f$, i.e, $f(h) = 0$. By proposition 3.1, $f(gh) = f(g)$ for every $g \in G$ and so $T_n(g) = 0$, for every n. Hence, $K_f \subset KerT$. It would be interesting to know under what conditions we have that $K_f = KerT$.*

References

[1] M. Avez, Entropie des groupes de type fini, C. R. Acad. Sc. Paris, Series A, T. **275**, n.25, (1972) 1363-1366.

[2] M. Avez, In: A. Dold, B. Eckmann (Eds.), Théorie Ergodique, Actes des Journées Ergodiques, Rennes 1973/1974, Lectures Notes in Mathematics, **532**, Springer-Verlag, (1976) 35-49

[3] G. Bettencourt and S. Mendes, Homomorphisms to $\mathbb{R}$ of semidirect products: a dynamical construction, Appl. Math. Inf. Sci., **9.6** (2015) 1-7.

[4] D. Calegari, scl, MSJ Memoirs, **20**, 2009.

[5] A. Erschler and A. Karlsson, Homomorphisms to $\mathbb{R}$ constructed from random walks, Annales de l'Institut Fourier **60.6** (2010) 2095-2113.

[6] N. Harrison, Real length functions in groups, Trans. Amer. Math. Soc. 174 (1972), 77-106

[7] Roger Lyndon, Length functions in groups, Math. Scand. 12 (1963), 209234

[8] A. Karlsson and F. Ledrappier, Linear drift and Poisson boundary for random walks, Pure Appl. Math. Q. **3** no. 4, part 1, (2007) 1027-1036.

[9] V. Kaimanovich and A. Vershik, Random Walks on Discrete Groups: Boundary and Entropy, Ann. Probab. Volume **11**, No. 3, (1983) 457-490.

[10] H. Kesten, Symmetric random walks on groups, Trans. Amer. Math. Soc. 92, (1959), 336-354.

[11] G. Pólya, Uber eine Aufgabe der Wahrscheinlichkeitstheorie betreffend dieIrrfahrt im Straßennetz, Math Ann. 84 (1921), 149-160.

[12] A. Vershik, Numerical characteristic of the groups and their relations, Zapiski Nauchn. Semin. POMI **256**, (1999) 5-18. English translation: J. Math. Sci. **107**, No. 5, (2001) 4147-4156.

Sudden transition from equilibrium stability to chaotic dynamics in a cautious tâtonnement model

I. Foroni[1], A. Avellone[1] and A. Panchuk[2]

E-mail: ilaria.foroni@unimib.it, alessandro.avellone@unimib.it, anastasiia.panchuk@gmail.com
[1] Department of Statistica e Metodi Quantitativi University of Milano-Bicocca (Italy)
[2] Institute of Mathematics, NAS of Ukraine, 3 Tereshchenkivska str., 01601 Kyiv (Ukraine)

Abstract. Discrete time price adjustment processes may fail to converge and may exhibit periodic or even chaotic behavior. To avoid large price changes, a version of the discrete time tâtonnement process for reaching an equilibrium in a pure exchange economy based on a *cautious* updating of the prices has been proposed two decades ago. This modification leads to a one dimensional bimodal piecewise smooth map, for which we show analytically that degenerate bifurcations and border collision bifurcations play a fundamental role for the asymptotic behavior of the model.

1. Introduction

In [12] the author observed that while habitually the eonomic models used to describe the discrete time tâtonnement process allowed very large price changes, "in reality price changes are usually restricted in a given period and are adjusted to substantial market disequilibria only over time" (p.1 in [12]). To overcome the shortcoming he suggested the assumption of a *cautious* process of price adjustment, that is, he proposed introducing a bound on the percentage price changes in function of the excess of the demand. As an example, he inserted the *cautious* mechanism in the simplest case of pure exchange economy introduced in the work of Day and Pianigiani ([2]). The analytical description of the model resulted expressed by a bimodal 1D continuous piecewise linear map which showed the possibility of erratic dynamics even in presence of the cautious adjustment. In the following years, the model has been extensively studied by many authors in consideration of its economic implications. Nonetheless, from the mathematical point of view the map describing the original model has not yet been completely explained, mainly because of the fact that the map is not smooth. To our knowledge, the bifurcation structure of the parameter space of a *generic* 1D piecewise linear bimodal map has been described to a large extent [9, 10]. Unlike the skew tent map, the piecewise linear bimodal map has *two* border points, which implies more tangled bifurcation structure of its parameter space. However, for certain parameter values the bimodal map (or its suitable iterate) admits reduction either to a skew tent map or to a discontinuous map defined in two partitions. The results stated in these two references are exploited in the present paper in order to describe asymptotic behavior of the map which models a cautious tâtonnement process. This paper is an abridged version of [3] to which we would like to refer interested readers for analytical proofs and details.

2. The model

The model framework considers a pure exchange economy with 2 individuals, which we label with an index i ($i = 1, 2$), and 2 commodities. The commodities have to be chosen in amounts y and w, respectively, and we assume that individuals' preferences are characterized by the following Cobb-Douglas utility functions

$$U^1(y, w) = y^{\alpha} w^{1-\alpha}, \tag{1}$$

$$U^2(y, w) = y^{\beta} w^{1-\beta} \tag{2}$$

where $0 < \alpha, \beta < 1$. Moreover, we consider that individual 1 is endowed with a quantity A of the first good while individual 2 is endowed with a quantity B of the second good. We indicate with p the unitary price of the first good while q denote the unitary price of the second good. Thus, we may express the budget constraint of consumer 1 as

$$py + qw \leq pA \tag{3}$$

and the budget constraint of consumer 2 as

$$py + qw \leq qB. \tag{4}$$

Let us suppose that prices are normalized so that the price q of the second good is set to a constant and only the price p of the first is adjusted, then maximizing (1) subject to (3) and (2) subject to (4) we obtain the following total excess of demand for the first good

$$z(p) = -(1-\alpha)A + \frac{\beta q B}{p}. \tag{5}$$

To introduce the tâtonnement process, we have to substitute p_t for p in (5) and then consider the difference equation

$$p_{t+1} = \max\{0, g(z(p_t))\}, \tag{6}$$

where $g(z(p_t))$ expresses the price adjustment process as a function of the excess of demand $z(p_t)$. In particular, choosing the relative price adjustment mechanism

$$g(z(p_t)) = p_t(1 + \lambda z(p_t)), \tag{7}$$

where λ is a positive constant representing the velocity of adjustment, expression (6) becomes

$$p_{t+1} = \max\{0, p_t(1 + \lambda z(p_t))\}. \tag{8}$$

To obtain exactly the model proposed in [12], we have to assume the hypothesis that prices are adjusted *cautiously* meaning that the amount of the percentage price changes is supposed to be bounded. Analytically this assumption can be expressed adding the restrictions

$$1 - r \leq \frac{g(z(p_t))}{p_t} \leq 1 + r, \tag{9}$$

where $0 < r < 1$ denotes the maximal rate of price change. Thus, by substituting (9) in (6) and after some manipulations, the price adjustment mechanism describing the evolution of the model is defined by the following difference equation

$$p_{t+1} = \max\{(1-r)p_t, \min\{(1+r)p_t, (\lambda\beta qB + (1 - \lambda(1-\alpha)A)p_t)\}\}. \tag{10}$$

By assuming the condition $\beta = 1-\alpha$ as in [12] we can rewrite (10) as a 1D continuous piecewise linear map, $p_{t+1} = g(p_t)$, with $g : [0,\infty) \to [0,\infty)$ defined by three linear functions $g_{\mathcal{L}}$, $g_{\mathcal{M}}$, and $g_{\mathcal{R}}$ as follows

$$g : p \mapsto g(p) = \begin{cases} g_{\mathcal{L}}(p) = (1+r)p & \text{for} \quad 0 \le p \le \frac{\varepsilon\gamma}{A(r+\varepsilon)}, \\ g_{\mathcal{M}}(p) = \frac{\varepsilon\gamma}{A} + (1-\varepsilon)p & \text{for} \quad \frac{\varepsilon\gamma}{A(r+\varepsilon)} < p \le \frac{\varepsilon\gamma}{A(\varepsilon-r)}, \\ g_{\mathcal{R}}(p) = (1-r)p & \text{for} \quad p > \frac{\varepsilon\gamma}{A(\varepsilon-r)}, \end{cases} \tag{11}$$

where r, ε, A, and γ are real parameters such that

$$0 < r < 1, \quad \varepsilon \equiv \lambda A\beta > r, \quad A > 0, \quad \gamma = qB > 0. \tag{12}$$

It can be proved (see [3] for details) that the map g is topologically conjugate to the following reduced map

$$f : x \to f(x) = \begin{cases} f_{\mathcal{L}}(x) = a_{\mathcal{L}}x + \mu_{\mathcal{L}} = (1+r)x, & 0 \le x \le x_{\mathcal{L}}, \\ f_{\mathcal{M}}(x) = a_{\mathcal{M}}x + \mu_{\mathcal{M}} = (1-\varepsilon)x + \varepsilon, & x_{\mathcal{L}} < x \le x_{\mathcal{R}}, \\ f_{\mathcal{R}}(x) = a_{\mathcal{R}}x + \mu_{\mathcal{R}} = (1-r)x, & x > x_{\mathcal{R}}, \end{cases} \tag{13}$$

where

$$a_{\mathcal{L}} = 1+r, \quad a_{\mathcal{M}} = 1-\varepsilon, \quad a_{\mathcal{R}} = 1-r \qquad \text{and} \qquad \mu_{\mathcal{L}} = \mu_{\mathcal{R}} = 0, \quad \mu_{\mathcal{M}} = \varepsilon$$

are the slopes and the offsets, respectively, while

$$x_{\mathcal{L}} = \frac{\varepsilon}{\varepsilon + r}, \quad x_{\mathcal{R}} = \frac{\varepsilon}{\varepsilon - r} \tag{14}$$

are the border points. Therefore, asymptotic dynamics of f is qualitatively the same as asymptotic dynamics of g, while the factor $\frac{\gamma}{A}$ only influences the scale. As a consequence, below we restrict our analysis to the (r,ε) parameter plane. The map f in (13) is bimodal as the branches $f_{\mathcal{L}}$ and $f_{\mathcal{R}}$ are monotonically increasing while the middle one, $f_{\mathcal{M}}$, is monotonically decreasing. By definition the border points $x_{\mathcal{L}}$ and $x_{\mathcal{R}}$ are both singular points for f. Moreover, f has a local maximum in $x_{\mathcal{L}}$, and a local minimum in $x_{\mathcal{R}}$, which are denoted as

$$x_M \stackrel{\text{df}}{=} f(x_{\mathcal{L}}) = \frac{\varepsilon(r+1)}{\varepsilon + r} \qquad x_m \stackrel{\text{df}}{=} f(x_{\mathcal{R}}) = \frac{\varepsilon(1-r)}{\varepsilon - r}. \tag{15}$$

3. Bifurcation structures in the parameter space

In this section, we begin the analysis of the model (10) introducing some of the properties of the map f which describes the price evolution of the first commodity in the economic system. We will show that only two kind of dynamics are allowed since it will be proved that, depending on the values of the parameters, the price or converges to a globally stable equilibrium or follows a chaotic dynamics.

3.1. Preliminary observations, feasible domain and absorbing intervals

The region D of definition of the map f is confined by three boundaries:

$$\delta_{ND} = \{(r,\varepsilon) : \varepsilon = r\}, \quad \delta_{r1} = \{(r,\varepsilon) : r = 1\}, \quad \delta_{r0} = \{(r,\varepsilon) : r = 0\}, \tag{16}$$

that is,

$$D = \{(r,\varepsilon) \,:\, \varepsilon > r,\ 0 < r < 1\}. \tag{17}$$

Journal of Physics: Conference Series **692** (2016) 012005 doi:10.1088/1742-6596/692/1/012005

Moreover, the map (13) has a fixed point x^0 in the origin which is always repelling. Due to this reason we restrict our analysis to the interval $(0,\infty)$ which is divided into the three *partitions*, $I_{\mathcal{L}} = (0, x_{\mathcal{L}}]$, $I_{\mathcal{M}} = (x_{\mathcal{L}}, x_{\mathcal{R}}]$, and $I_{\mathcal{R}} = (x_{\mathcal{R}}, \infty)$, on which the functions $f_{\mathcal{L}}$, $f_{\mathcal{M}}$, and $f_{\mathcal{R}}$ are defined. For parameter values belonging to the feasible domain D given in (17) the map f always admits an absorbing interval J which is globally attracting. The existence of such an interval is a very important feature for the system, because *all* orbits are trapped by the absorbing set, and what happens in this set determines the long-run dynamics of the model. These preliminary results prove an important economic property of the model since we may affirm that for the values of the parameters considered it is not possible that price p becomes zero or diverges. Moreover, the interval in which the price varies may exclude big jumps if the relative price adjustment is restricted by a maximal rate of increase or decrease. The investigation of the absorbing interval of f allows us to identify which ones of the branches defined in (13) are involved in the asymptotic behavior of f. Thus, D may be divided into three sub-regions denoted by $D_{\mathcal{M}}$, $D_{\mathcal{LM}}$, and $D_{\mathcal{LMR}}$. These regions distinguish situations in which different numbers of branches of f are involved in its asymptotic dynamics as reflected by the subscripts. That is, asymptotic behavior of the map f involves only the middle branch ($D_{\mathcal{M}}$), the left and the middle branches ($D_{\mathcal{LM}}$), or all three branches ($D_{\mathcal{LMR}}$). According to this, we denote the absorbing interval of f by $J_{\mathcal{M}}$, $J_{\mathcal{LM}}$, or $J_{\mathcal{LMR}}$ when we consider parameter values in $D_{\mathcal{M}}$, $D_{\mathcal{LM}}$ or $D_{\mathcal{LMR}}$, respectively.

3.2. A stable fixed point: region $D_{\mathcal{M}}$

Apart from the repelling origin x^0, the map f always has another fixed point $x^* = 1$ which may be stable or unstable. By definition x^* is stable when the related slope $|f'(x^*)| = |a_{\mathcal{M}}| < 1$, which clearly holds whenever $\varepsilon < 2$. The line

$$\delta_{DFB} = \{(r, \varepsilon) \,:\, \varepsilon = \varepsilon_{DFB} = 2\} \tag{18}$$

defines a *degenerate flip bifurcation* (DFB) [11] boundary in the parameter plane of f. Thus, the point x^* is stable for the parameter values belonging to region $D_{\mathcal{M}} = \{(r, \varepsilon) : 0 < r < 1,\ r < \varepsilon < 2\}$ confined by three feasible domain boundaries δ_{ND}, δ_{r0}, δ_{r1}, and the DFB boundary δ_{DFB}. Moreover, since the slope of the left branch is $a_{\mathcal{L}} > 1$ and the slope of the right branch is $a_{\mathcal{R}} < 1$, any orbit of f enters the middle partition $I_{\mathcal{M}}$ in a finite number of iterations, and hence, the point x^* is globally attracting. Thus, as $\varepsilon \equiv \lambda A \beta$, we can observe that the constant elasticity of substitution of the Cobb-Douglas utility functions, the endowment of the first good and the velocity of the price adjustment mechanism are the parameters influencing the dynamics of the model together with the maximal rate of price change. Furthermore, we can affirm that the values of the parameters λ, A and β may play a crucial role in preserving the stability of the equilibrium price whatever the value of r is.

3.3. Reduction to a skew tent map: region $D_{\mathcal{LM}}$

Immediately after the DFB occurring at δ_{DFB} the map f has an absorbing interval $J_{\mathcal{LM}} = \left[f^2(x_{\mathcal{L}}), f(x_{\mathcal{L}})\right] = J^1_{\mathcal{LM}} \cup J^2_{\mathcal{LM}}$ with

$$J^1_{\mathcal{LM}} = \left[f^2(x_{\mathcal{L}}), x_{\mathcal{L}}\right] = \left[\frac{\varepsilon\,(2r - r\varepsilon + 1)}{r + \varepsilon}, \frac{\varepsilon}{r + \varepsilon}\right] \subset I_{\mathcal{L}}$$

and

$$J^2_{\mathcal{LM}} = (x_{\mathcal{L}}, f(x_{\mathcal{L}})] = \left(\frac{\varepsilon}{r + \varepsilon}, \frac{\varepsilon(1 + r)}{r + \varepsilon}\right] \subset I_{\mathcal{M}}.$$

Thus, the map f restricted to $J_{\mathcal{LM}}$ is defined as

$$f_{J_{\mathcal{LM}}} \,:\, x \mapsto f_{J_{\mathcal{LM}}}(x) = \begin{cases} f_{\mathcal{L}}(x) = (1 + r)\,x, & \frac{\varepsilon(2r - r\varepsilon + 1)}{r+\varepsilon} \le x \le \frac{\varepsilon}{r+\varepsilon}, \\ f_{\mathcal{M}}(x) = \varepsilon + (1 - \varepsilon)\,x, & \frac{\varepsilon}{r+\varepsilon} < x \le \frac{\varepsilon(1+r)}{r+\varepsilon}. \end{cases} \tag{19}$$

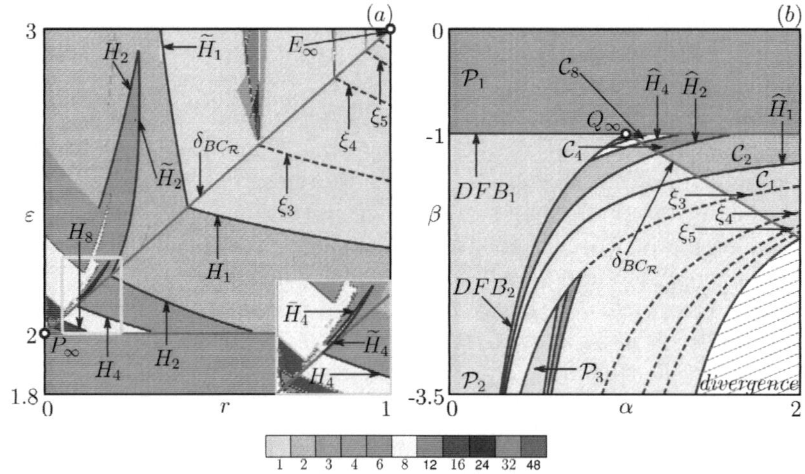

Figure 1. (a) Parameter plane of the skew tent map $\hat{f}$. The region corresponding to $D_{\mathcal{LM}}$ is outlined red. H_{2^m} and $\widetilde{H}_{2^m}$ are the curves (accumulating at the point P_∞) where the first homoclinic bifurcation of the repelling cycles $\mathcal{O}_{2^m}$ occur. (b) Skew tent structure inside the region $D_{\mathcal{LM}}$ for the map f and its extension to the region $D_{\mathcal{LMR}}$. $\widehat{H}_{2^m}$ indicates the boundaries (accumulating at the point Q_∞) corresponding to the first homoclinic bifurcation of the repelling cycle $\mathcal{O}_{2^m}$. In both panels ξ_q designate the fold BCB leading to the appearance of the two unstable cycles $\mathcal{O}_q$ and $\widehat{\mathcal{O}}_q$.

The map (19) is applicable while the absorbing interval $J_{\mathcal{LM}}$ remains in the left and the middle partitions, that is, until $f(x_{\mathcal{L}}) = x_{\mathcal{R}}$ is fulfilled and the right endpoint of $J_{\mathcal{LM}}$ coincides with $x_{\mathcal{R}}$. This contact happens when ε and r belong to the line $\delta_{BC_{\mathcal{R}}} = \{(r, \varepsilon) : \varepsilon = r + 2\}$, so that $\delta_{BC_{\mathcal{R}}}$ is the boundary between the regions $D_{\mathcal{LM}}$ and $D_{\mathcal{LMR}}$. In such a way the region $D_{\mathcal{LM}}$ is defined as $D_{\mathcal{LM}} = \{(r, \varepsilon) : 0 < r < 1,\ 2 < \varepsilon < r + 2\}$. The 2D bifurcation diagram shown in Fig. 1(a) displays how the bifurcation structure changes when the parameter point moves from $D_{\mathcal{M}}$ to $D_{\mathcal{LM}}$. In the diagram, gray color indicates convergence of the orbits of the map f to the fixed point x^*, while the other colors denote regions associated with cyclic chaotic intervals that is chaotic attractors made up of different bands (correspondence between colors and numbers of intervals is indicated in the color bar). As the 2D bifurcation diagram in Fig. 1(a) suggests, the main feature concerning the asymptotic dynamics of (13) in $D_1 \cup D_2$ is that crossing the straight line δ_{DFB} (18) leads to the abrupt transition from the global stability of the fixed point x^* in region $D_{\mathcal{M}}$ to chaos which characterizes the whole area $D_{\mathcal{LM}}$. Thus, the introduction of the cautious price updating in the tâtonnement model causes a sudden change from the stable equilibrium price to the chaotic regime which is rather exceptional economic phenomenon. Not only, but as it will be proved in the following analysis, the complete absence of periodic attracting cycles is a feature of this version of the model that distinguishes it from its classical formulation. As we will show, two characteristics of (13) mainly contribute to the peculiar bifurcation structure of the set $D_{\mathcal{M}} \cup D_{\mathcal{LM}}$ in the (r, ε)-plane: first, the change in stability of x^* when we pass from $D_{\mathcal{M}}$ to $D_{\mathcal{LM}}$, and second, the unimodal shape of $f_{J_{\mathcal{LM}}}$ in (19). The latter fact allows us to assert that the map $f_{J_{\mathcal{LM}}}$ is topologically conjugate to the skew tent map in the normal form

$$\hat{f} : x \to \hat{f}(x) = \begin{cases} \hat{f}_{\mathcal{A}}(x) = \alpha x + 1, & x \le 0, \\ \hat{f}_{\mathcal{B}}(x) = \beta x + 1, & x > 0, \end{cases} \tag{20}$$

with

$$\alpha = a_{\mathcal{L}} = 1 + r, \quad \beta = a_{\mathcal{M}} = 1 - \varepsilon.$$

As it is known, the skew tent map has been widely studied by many authors and for its normal form $\hat{f}$ all the possible kinds of border collision bifurcations are classified (see [4, 8, 6, 7] to cite a few, and also [11] for a detailed description and references therein). From the topological conjugacy of (19) and (20) it follows that if we refer to the 2D bifurcation diagram of the skew tent map $\hat{f}$ we may deduce the bifurcation structure of the region $D_{\mathcal{LM}}$ of the map f. In particular, we can explain the meaning of the curves that divide $D_{\mathcal{LM}}$ into sub-regions as it is shown in the 2D bifurcation diagram of Fig. 1(a). For that we plot in Fig. 1(b) a bifurcation diagram in the (α, β) parameter plane of $\hat{f}$ in order to compare it with the panel (a) (the red line marks the parameter region which corresponds to $D_{\mathcal{LM}}$). As it is immediately seen, the asymptotic dynamics of the skew tent map $\hat{f}$ for the related parameter values corresponds to cyclic chaotic intervals $\mathcal{Q}_{2^m}$, $m = 0, 1, \ldots$, whose regions are denoted as $\mathcal{C}_{2^m}$. It can be also evidenced that after the DFB of the fixed point x^* there holds $|\alpha\beta| > 1$. Hence, the cycle of period 2 which is born due to the bifurcation is unstable, and therefore the DFB of x^* leads to a chaotic regime. For the skew tent map $\hat{f}$ it is known (see, *e. g.*, [11]) that any two contiguous regions $\mathcal{C}_{2^m}$ and $\mathcal{C}_{2^{m+1}}$ are separated from each other by the boundary corresponding to the first homoclinic bifurcation of the repelling cycle $\mathcal{O}_{2^m}$ of period 2^m. Analytically the homoclinic bifurcation boundary is expressed in general case as

$$\widehat{H}_{2^m} = \left\{ (\alpha, \beta) \ : \ \alpha^{2\delta_m} \beta^{2\delta_{m+1}} + \left(\frac{\alpha}{\beta} \right)^{(-1)^{m+1}} - 1 = 0 \right\}, \tag{21}$$

where

$$\delta_m = \frac{2^m - (-1)^m}{3}, \quad m \geq 0. \tag{22}$$

All curves of the form (21) accumulate at the point $Q_\infty(\alpha, \beta) = (1, -1)$ when $m \to \infty$. The expression (21) is easily mapped to the parameter plane (r, ε) of the original map f as follows

$$H_{2^m} = \left\{ (r, \varepsilon) \in D_{\mathcal{LM}} : \ (1 + r)^{2\delta_m} (1 - \varepsilon)^{2\delta_{m+1}} + \left(\frac{1 + r}{1 - \varepsilon} \right)^{(-1)^{m+1}} - 1 = 0 \right\}. \tag{23}$$

Moreover, the accumulation point Q_∞ is translated to $P_\infty(r, \varepsilon) = (0, 2)$. In Fig. 1(a) the homoclinic bifurcation curves H_{2^m}, $m = 0, \ldots, 4$, are numerically obtained from (23). Note that the other curves for $m > 4$ also exist, but they are not observable in the presented scale, and these curves accumulate at the point P_∞. It is also easy to check that all the homoclinic curves H_{2^m} for any integer $m \geq 1$ intersect the line δ_{DFB}. So that, depending on the value of parameter r the DFB of the fixed point x^* occurring at $\varepsilon = 2$ can lead to cyclical chaotic intervals $\mathcal{Q}_{2^m}$ for any integer $m \geq 1$. This is not true for the curve H_1 which is asymptotic to the DFB line $\varepsilon = 2$ meaning that we cannot have direct transition from a stable equilibrium point to one chaotic interval $\mathcal{Q}_1$. Figs. 1 (a) and (b) also display several other bifurcation curves, denoted as ξ_q, $q \geq 3$, plotted by dashed lines. Each such boundary corresponds to the fold BCB leading to the appearance of the two cycles $\mathcal{O}_q$ and $\widehat{\mathcal{O}}_q$ of the same period q. The cycle $\mathcal{O}_q$ is the so-called *basic* cycle having a single point in $I_{\mathcal{M}}$ and all other $q - 1$ points in $I_{\mathcal{L}}$. The other cycle $\widehat{\mathcal{O}}_q$ is called *complementary* and it has two points in $I_{\mathcal{M}}$. For parameter values belonging to $D_{\mathcal{LM}}$ both cycles, $\mathcal{O}_q$ and $\widehat{\mathcal{O}}_q$, appear unstable. In the (r, ε)-parameter plane of the map f the fold BCB curve at which $\mathcal{O}_q$ and $\widehat{\mathcal{O}}_q$ appear is given by

$$\xi_q = \left\{ (r, \varepsilon) \ : \ 1 - \varepsilon = -\frac{1 - (1 + r)^{q-1}}{r(1 + r)^{q-2}} \right\}, \quad q \geq 3, \tag{24}$$

then the two cycles exist in the region above the curve ξ_q. It can be proved that $\xi_q \cap D_{\mathcal{LM}} \neq \emptyset$ for any $q \geq 3$, and all these curves accumulate at the point $E_\infty(r, \varepsilon) = (1, 3)$ for $q \to \infty$. As one can see in Fig. 1(a), the intersection point of each ξ_q and the border $\delta_{BC_\mathcal{R}}$ is particular. In fact, as we shall see, each such point is the issuing point of other bifurcation curves located in region $D_{\mathcal{LMR}}$ and associated with cyclic chaotic intervals of the map f. These chaotic intervals are related to period adding structure.

3.4. All three partitions: region $D_{\mathcal{LMR}}$

The transition from region $D_{\mathcal{LM}}$ to region $D_{\mathcal{LMR}}$ occurs when $(r, \varepsilon) \in \delta_{BC_\mathcal{R}}$. Hence, the region $D_{\mathcal{LMR}}$ is defined as

$$D_{\mathcal{LMR}} = \{(r, \varepsilon) \, : \, 0 < r < 1, \; \varepsilon > r + 2\}. \tag{25}$$

For parameter values belonging to $D_{\mathcal{LMR}}$ the expression of the absorbing interval is

$$J_{\mathcal{LMR}} = [f(x_\mathcal{L}), f(x_\mathcal{R})] = \left[\frac{(1-r)\varepsilon}{\varepsilon - r}, \frac{(1+r)\varepsilon}{\varepsilon + r}\right]$$

and, as it can be easily checked, $J_{\mathcal{LMR}}$ ranges across all the partitions $I_\mathcal{L}$, $I_\mathcal{M}$, and $I_\mathcal{R}$. As a consequence, the definition of the map f in the region $D_{\mathcal{LMR}}$ involves all three branches. Preliminary understanding of the bifurcation structure of the region $D_{\mathcal{LMR}}$ can be obtained from Fig. 1(a). As one can see, when the parameter point moves from $D_{\mathcal{LM}}$ to $D_{\mathcal{LMR}}$ the regions associated with chaotic dynamics existing in $D_{\mathcal{LM}}$ extend as well to $D_{\mathcal{LMR}}$. In this section we describe briefly why the related boundaries change and find analytic expressions for some of them. At first, we consider the boundary $H_1 \subset D_{\mathcal{LM}}$ related to the first homoclinic bifurcation of the fixed point x^*, and let the parameter point p move along this boundary. When the parameter point crosses $\delta_{BC_\mathcal{R}}$ and enters the region $D_{\mathcal{LMR}}$, the condition for the first homoclinic bifurcation of the fixed point x^* defines the new curve

$$\widetilde{H}_1 = \left\{(r, \varepsilon) \, : \, \varepsilon = \frac{1}{r}\right\}. \tag{26}$$

Similarly, all the other homoclinic bifurcation curves $H_{2^m} \subset D_{\mathcal{LM}}$, $m \geq 1$, have their extensions inside $D_{\mathcal{LMR}}$. As it can be seen in Fig. 1(a), the regions related to cyclic chaotic intervals belonging to $D_{\mathcal{LMR}}$ have more complicated structure than those belonging to $D_{\mathcal{LM}}$. Moreover, the higher the number of chaotic intervals, the higher the number of bifurcation curves bounding the related parameter regions. To explain the bifurcation structure of the region $D_{\mathcal{LMR}}$ when the parameter point crosses the curve $\widetilde{H}_1$, we have to consider the *period adding (PA) structure* of the map f. It is known for a generic bimodal piecewise linear map that even if its absorbing interval spreads over all three partitions the map still may have periodic orbits whose points are located only in the two outermost partitions $I_\mathcal{L}$ and $I_\mathcal{R}$. Existence conditions for such cycles can be derived by means of a discontinuous map defined in two partitions, as shown, for instance, in [9]. In particular, periodicity regions related to these cycles are arranged in the parameter space into a *PA structure*, known also as *Arnold tongues* or *mode-locking tongues*. However, that for the map of the form (13) the PA structure is highly degenerate and consists only of a set of lines defined by $r =$ const. To obtain the boundaries of regions constituting the PA structure (PA regions, for short) in the parameter plane of the map f we use the expressions for such boundaries known for a generic bimodal piecewise linear map. Let us start with regions of the so-called first complexity level (following Leonov [5]) associated with q-cycles having either a single point in $I_\mathcal{L}$ and all other $q-1$ points in $I_\mathcal{R}$ (type I), or vice verse (type II). The BCB

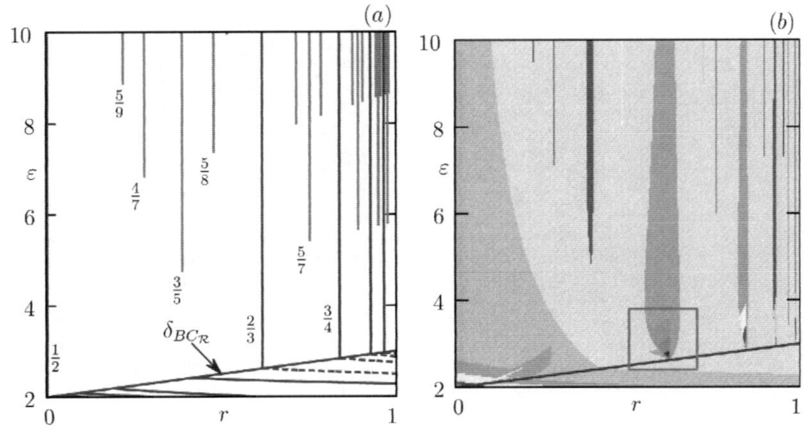

Figure 2. (a) Degenerate PA structures in the (r, ε) parameter plane of the map f. Blue lines correspond to first complexity level cycles, red lines correspond to those of the second complexity level. (b) Numerically calculated chaoticity regions around PA structure.

boundaries of regions related to the type I cycles are defined in general form as

$$\begin{aligned} &\text{BC with } x_\mathcal{L}: \quad \mu_\mathcal{L} = -\frac{1-a_\mathcal{R}^{q-1}}{(1-a_\mathcal{R})a_\mathcal{R}^{q-1}}\mu_\mathcal{R} + \frac{1-a_\mathcal{R}^{q-1}a_\mathcal{L}}{a_\mathcal{R}^{q-1}}x_\mathcal{L}, \\ &\text{BC with } x_\mathcal{R}: \quad \mu_\mathcal{L} = -\left(a_\mathcal{L} + \frac{1-a_\mathcal{R}^{q-2}}{(1-a_\mathcal{R})a_\mathcal{R}^{q-2}}\right)\mu_\mathcal{R} + \frac{1-a_\mathcal{R}^{q-1}a_\mathcal{L}}{a_\mathcal{R}^{q-2}}x_\mathcal{R}. \end{aligned} \tag{27}$$

The BCB boundaries of the type II cycles are obtained from (27) by interchanging the indices $\mathcal{L}$ and $\mathcal{R}$ (for detail see [1, 9]). Due to the fact that $\mu_\mathcal{L} = \mu_\mathcal{R} = 0$ for f, the expressions for BCB boundaries of PA regions related to the type I cycles are clearly simplified to

$$\text{BC with } x_\mathcal{L}: \frac{(1-a_\mathcal{R}^{q-1}a_\mathcal{L})x_\mathcal{L}}{a_\mathcal{R}^{q-1}} = 0, \quad \text{BC with } x_\mathcal{R}: \frac{(1-a_\mathcal{R}^{q-1}a_\mathcal{L})x_\mathcal{R}}{a_\mathcal{R}^{q-2}} = 0. \tag{28}$$

From (28) it is obviously deduced that for a type I cycle both BCBs, with $x_\mathcal{L}$ and $x_\mathcal{R}$, occur at the same time, namely, when $a_\mathcal{R}^{q-1}a_\mathcal{L} = 1$. Similarly, for a type II cycle both BCBs occur when $a_\mathcal{L}^{q-1}a_\mathcal{R} = 1$. Hence, for the map f any cycle related to PA region of the first complexity level exists only when its multiplier equals to one. Taking into account that $a_\mathcal{L} = 1 + r$, $a_\mathcal{R} = 1 - r$ we obtain that any PA region of the first complexity level degenerates to a line $r = \bar{r}_{q,1}$ with $\bar{r}_{q,1}$ depending on q. The other regions belonging to PA structure (of the second and higher complexity levels) are degenerate as well. Having (28), we derive that for the map f *any* PA-cycle of *any* complexity level exists only when its multiplier equals to one, that is, for $a_\mathcal{L}^l a_\mathcal{R}^{q-l} = 1$. Then the related PA region degenerates to a line $r = \bar{r}_{q,l}$ with $\bar{r}_{q,l}$ depending on the period of the cycle and the number of points in $I_\mathcal{L}$. In Fig. 2(a) we plot several lines corresponding to PA regions of the first (blue) and second (red) complexity level. Although for the map f the PA structure is highly degenerate, it still plays an essential role in the bifurcation structure in the parameter plane of f. In particular, the set of lines composing PA structure serves as a skeleton for the regions in the parameter plane of f related to the cycles of chaotic intervals shown in Fig. 2(b). In that figure the area painted light-blue corresponds to a single chaotic interval enclosing both border points, while the other colors are related to different number of chaotic intervals (as indicated in the color bar) which can enclose only one border point ($x_\mathcal{L}$

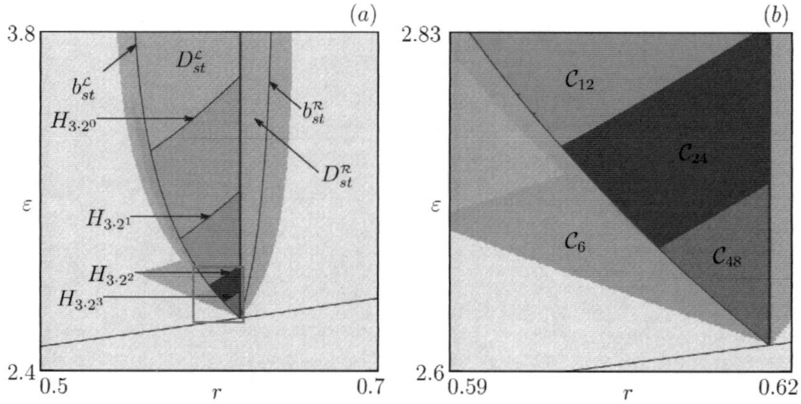

Figure 3. Chaoticity regions around the line $(1+r)^2(1-r) = 1$ corresponding to $\mathcal{O}_{\mathcal{RL}^2}$. (a) In regions $D^{\mathcal{L}}_{st}$ and $D^{\mathcal{R}}_{st}$ the map f^3 in the neighborhood of the border point $x_{\mathcal{L}}$ is topologically conjugate to a skew tent map. Curves indicated by $H_{3\cdot2^m}$ are boundaries separating regions characterized by the existence of $3\cdot2^m$ pieces chaotic attractors. (b) Chaotic regions related to $3\cdot2^m$ pieces chaotic attractors are indicated by $\mathcal{C}_{3\cdot2^m}$.

or $x_{\mathcal{R}}$) or both of them. Comparing Figs. 2(a) and (b) one can see that the regions related to more than one chaotic interval are located in the neighborhood of the lines composing the PA structure. Some of the boundaries of these regions can be found analytically as described, for instance, in [10]. We concentrate below only on those boundaries which intersect with the lines composing the PA structure, and derive expressions for the region boundaries which issue from the line $r = \bar{r}_{3,2}$ associated with the cycle of period 3. In a similar way, such boundaries can be obtained for the regions related to any cycle in the PA structure. Consider $r = \bar{r}_{3,2}$, so that a cycle of period 3 having 2 points in $I_{\mathcal{L}}$ exists. First let $r < \bar{r}_{3,2}$. Following [10], we consider the map f^3 in the neighborhood of the border point $x_{\mathcal{L}}$. For the values of r sufficiently close to $\bar{r}_{3,2}$ the map f^3 in the neighborhood of $x_{\mathcal{L}}$ is topologically conjugate to a skew tent map. More precisely, such a conjugacy holds until

$$f^3(x_{\mathcal{L}}) \equiv f_{\mathcal{L}} \circ f_{\mathcal{R}} \circ f_{\mathcal{L}}(x_{\mathcal{L}}) < \bar{x}, \tag{29}$$

where $\bar{x}$ is a preimage of $x_{\mathcal{R}}$ such that $f_{\mathcal{M}}(\bar{x}) = x_{\mathcal{R}}$. The equality in (29) defines a curve in the parameter plane of f:

$$\begin{aligned} b^{\mathcal{L}}_{st} &= \{(r,\varepsilon) : f_{\mathcal{L}} \circ f_{\mathcal{R}} \circ f_{\mathcal{L}} \circ f_{\mathcal{M}}(x_{\mathcal{L}}) = x_{\mathcal{R}}\} \\ &= \left\{(r,\varepsilon)\ :\ \frac{(1-\varepsilon)(1+r)^2(1-r)}{\varepsilon+r} + 1 < \frac{\varepsilon}{\varepsilon-r}\right\}. \end{aligned} \tag{30}$$

In other words, for any set of parameter values belonging to the region $D^{\mathcal{L}}_{st}$ confined by $\mathfrak{L}_{3,2} = \{(r,\varepsilon)\ :\ r = \bar{r}_{3,2},\ \varepsilon \geq \bar{\varepsilon}_{3,2}\}$ and $b^{\mathcal{L}}_{st}$, the map f^3 in the neighborhood of the border point $x_{\mathcal{L}}$ is topologically conjugate to a skew tent map $\hat{f}$ given in (20) with $\alpha = (1+r)(1-r)^2$, $\beta = (1-r)^2(1-\varepsilon)$. The related boundaries of chaotic regions are then obtained from the expressions known for the skew tent map. In Fig. 3(a) we observe chaotic regions $\mathcal{C}_{3\cdot2^m}$, $m = 0, 4$. Now we turn to chaoticity regions located on the right-hand side of $\mathfrak{L}_{3,2}$, that is for $r > \bar{r}_{3,2}$. Similarly to the previous case, we consider the map f^3, but now in the neighborhood of the border point $x_{\mathcal{R}}$. The map f^3 in the neighborhood of $x_{\mathcal{R}}$ is topologically conjugate to a skew tent map $\hat{f}$ with $\alpha = (1+r)(1-r)^2$, $\beta = (1-r^2)(1-\varepsilon)$ as long as the inequality

$$f^3(x_{\mathcal{R}}) \equiv f_{\mathcal{R}} \circ f_{\mathcal{L}} \circ f_{\mathcal{L}}(x_{\mathcal{R}}) < \hat{x}, \tag{31}$$

holds with $f_{\mathcal{M}} \circ f_{\mathcal{L}}(\hat{x}) = x_{\mathcal{L}}$. Changing inequality to equality in (31) we get the curve

$$\begin{aligned} b_{st}^{\mathcal{R}} &= \{(r,\varepsilon) : f_{\mathcal{R}} \circ f_{\mathcal{L}} \circ f_{\mathcal{L}} \circ f_{\mathcal{M}} \circ f_{\mathcal{L}}(x_{\mathcal{R}}) = x_{\mathcal{L}}\} = \\ &= \{(r,\varepsilon) \ : \ \frac{(1-\varepsilon)(1+r)^3(1-r)}{\varepsilon - r} + 1 + r < \frac{\varepsilon}{\varepsilon + r}\}. \end{aligned} \quad (32)$$

Then for parameter values belonging to the region $D_{st}^{\mathcal{R}}$ confined by $\mathfrak{L}_{2,3}$ and the curve $b_{st}^{\mathcal{R}}$ the boundaries of chaotic regions are obtained from expressions known for the skew tent map. However, as it can be observed in Fig. 3(a) for $r > \bar{r}_{3,2}$ there is only a region $\mathcal{C}_3$ related to three pieces chaotic attractor. It can be proved that inside $D_{st}^{\mathcal{R}}$ no other chaotic regions appear.

4. Concluding remarks

In this paper we studied the map proposed in [12] to describe a cautious discrete time tâtonnement process in a pure exchange economy. The model is analytically expressed by a 1D piecewise linear bimodal map and the presence of two border points entails a complex bifurcation structure of its parameter space. Although the map has three branches it has been shown that for some parameter values its asymptotic dynamics involved only two of them. In these cases, results known for discontinuous maps or for the skew tent map have been applied. The main finding of our analyses can be resumed in the dichotomy which characterizes the dynamic behavior of the model. In fact, depending on the values of the parameters only one of the two following situations may occur: either the non trivial equilibrium point is globally stable or the orbits are chaotic. Moreover an abrupt transition characterizes the change from the first to the second regime. From an economic point of view, we may thus affirm that the introduction of the cautious process adjustment may reduce the amplitude of the price oscillations but may not eliminate the chaotic behavior typical of the discrete time tâtonnement process.

Acknowledgments

We thank the anonymous referees for their helpful comments. The work of I. Foroni and A. Panchuk has been done within the activities of the COST action IS1104 "The EU in the new economic complex geography: models, tools and policy evaluation" (www.gecomplexitycost.eu).

References

[1] Avrutin V, Schanz, M and Gardini L 2010 Calculation of bifurcation curves by map replacement *Int. J. Bifurcat. Chaos* **20** 3105-35
[2] Day RH and Pianigiani G 1991 Statistical dynamics and economics *J. Econ. Behav. Organ.* **16** 37-83
[3] Foroni I, Avellone A and Panchuk A 2015 Sudden transition from equilibrium stability to chaotic dynamics in a cautious tâtonnement model *Preprint* doi:10.1016/j.chaos.2015.05.013
[4] Ito S, Tanaka S and Nakada H 1979 On unimodal transformations and chaos II *Tokyo J. Math.* **2** 241-59
[5] Leonov N N 1959 Map of the line onto itself *Radiofisika* **3** 942-956
[6] Maistrenko Y L, Maistrenko V L and Chua L O 1993 Cycles of chaotic intervals in a time-delayed Chua's circuit, *Int. J. Bifurcat. Chaos* **3** 1557-72
[7] Maistrenko Y L, Maistrenko V L, Vikul, S I and Chua L O 1995 Bifurcations of attracting cycles from time-delayed Chua's circuit *Int. J. Bifurcat. Chaos* **5** 653-71
[8] Nusse HE and Yorke JA 1995 Border-collision bifurcation for piecewise smooth one-dimensional maps 1995 *Int. J. Bifurcat. Chaos* **5** 189-207
[9] Panchuk A, Sushko I, Schenke B and Avrutin V 2013 Bifurcation structure in a bimodal piecewise linear map: regular dynamics *Int. J. Bifurcat. Chaos* **23** 1330040.
[10] Panchuk A, Sushko I and Avrutin V 2015 Bifurcation structures in a bimodal piecewise linear map: chaotic dynamics *Int. J. Bifurcat. Chaos* **25** 1530006
[11] Sushko I and Gardini L 2010 Degenerate bifurcations and border collisions in piecewise smooth 1D and 2D maps *Int. J Bifurcat. Chaos* **20** 2045-70
[12] Weddepohl C 1995 A cautious price adjustment mechanism: chaotic behavior *J. Econ. Behav. Organ.* **27** 293-300

Journal of Physics: Conference Series **692** (2016) 012006 doi:10.1088/1742-6596/692/1/012006

Maps with vanishing denominator explained through applications in Economics

Fabio Tramontana
Department of Mathemtical Sciences, Mathematical Finance and Econometrics
Catholic University of Milan, Italy
E-mail: fabio.tramontana@unicatt.it

Abstract. We present the main results about plane maps with vanishing denominator by using as examples some economic applications where these results are used to investigate the global properties of the dynamic model. This kind of maps, originally brought to the attention of the researchers through their appearence in an economic application, are now well understood and the related results can now be applied to new models.

1. Introduction

Consider a plane map T of the following form:

$$T : \begin{cases} x' = G(x,y) \\ y' = F(x,y) = \dfrac{N(x,y)}{D(x,y)} \end{cases} \tag{1}$$

with $(x,y) \in \mathbb{R}^2$, $G(x,y)$, $N(x,y)$ and $D(x,y)$ are continuously differentiable functions defined on $\mathbb{R}^2$. The set of points where $D(x,y)$ vanishes is called *set of nondefinition* of the map T:

$$\delta_s = \left\{ (x,y) \in \mathbb{R}^2 \middle| \, D(x,y) = 0 \right\}, \tag{2}$$

which is assumed to be the union of smooth curves of the plane.

An interesting question is related to what happens to the image by T of an arc γ, transverse to δ_s and crossing it in correspondence of the point (x_0, y_0). If we parameterize the arc as $\gamma(\tau)$, such that $\gamma(0) = (x_0, y_0)$, we have that in general:

$$\lim_{\tau \to 0\pm} T(\gamma(\tau)) = (G(x_0, y_0), \infty),$$

with ∞ either equal to $+\infty$ or $-\infty$. In other words, the image $T(\gamma)$ is the union of two disjoint unbounded arcs asymptotic to the line of equation $x = G(x_0, y_0)$. This result is correct provided that in the point (x_0, y_0) only $D(x,y)$ vanishes, but what would happen if also $N(x,y)$ becomes 0 in correspondence of (x_0, y_0)?

This is exactly the situation that came out in an economic model that dates back to 1997, studied by Bischi and Naimzada (1997). They obtained quite particular structures of the basins of attraction, founding that these phenomena were related to the presence of points where a component of the map assumes the form 0/0.

Journal of Physics: Conference Series **692** (2016) 012006 doi:10.1088/1742-6596/692/1/012006

Moved by the necessity of solving the puzzle originated in the economic application, some authors (see Gardini & Bischi 1996, Bischi & Gardini 1997, Bischi et al. 1999, 2003, 2005, Gardini et al. 2007)[1] developed a theory that permits to explain not only the shape of the basins of attractions of the original model but also other interesting global phenomena arising in different fields of application such as Biology (Gu & Huang 2006, Gu 2009, Djellit et al. 2013) and Population Dynamics (Gu & Hao 2007). The theory developed to solve a particular problem has been extended and later applied to other models (see Bischi et al. 2001a for an early survey without reference to particular fields of application). So, it is interesting as in this case we can see in action the bidirectionality between theory and applications that characterize scientific research.

Recently, other economic models characterized by a map with vanishing denominator, have been studied by using the results of Bischi and collaborators. In this paper we summarize some of the most important theoretical results concerning this class of maps by using as examples the economic applications. In such a way we will also obtain a survey of the economic applications of maps with vanishing denominator.

2. Basic definitions

A point $Q = (x_0, y_0)$ such that $D(x_0, y_0) = N(x_0, y_0) = 0$ is called *focal point* if there exist smooth simple arcs $\gamma(\tau)$, with $\gamma(0) = Q$ and $\lim_{\tau \to 0} T(\gamma(\tau))$ is finite.

If we take different arcs $\gamma(\tau)$ through Q, the union of all the finite values of such a limit form the so-called *prefocal set* (or *curve*), denoted by δ_Q and belonging to $x = G(Q)$.

A focal point is called *simple* if:

$$\overline{N}_x \overline{D}_y - \overline{N}_y \overline{D}_x \neq 0, \tag{3}$$

where $\overline{N}_x = (\partial N / \partial x)(x_0, y_0)$ and the other partial derivatives are analoguous. Otherwise Q is called *non simple focal point*. Stated differently, a focal point is simple if it a simple root of the algebraic system made up by $N(x, y) = 0$ and $D(x, y) = 0$.

2.1. Main results and applications when the focal point is simple

One of the most important results related to the presence of a simple focal point is the following:

Proposition 1 *(Bischi et al., 1999). If Q is a simple focal point of a fractional rational map such as (1), then the following one-to-one relation between the slope m of an arc γ through Q and the point $(G(Q), y)$ in which $T(\gamma)$ crosses δ_Q exists:*

$$m \to (G(Q), y(m)),$$

with

$$y(m) = \frac{\overline{N}_x + m\overline{N}_y}{\overline{D}_x + m\overline{D}_y}$$

and

$$(F(Q), y) \to m(y)$$

with

$$m(y) = \frac{\overline{D}_x y - \overline{N}_x}{\overline{N}_y - \overline{D}_y y}.$$

[1] Some earlier work where maps with vanishing denominator were also considered are Mira (1981, 1996).

In other words T maps different arcs γ_j through Q with different slopes m_j into bounded arcs $T(\gamma_j)$ crossing the prefocal curve δ_Q in different points $(G(Q), y(m_j))$.

Even more interesting is to look at the action of the inverse map T^{-1} (or one of the inverses if the map is noninvertible). If we have an arc ω moving towards the prefocal curve δ_Q, its preimage $\omega^{-1} = T^{-1}(\omega)$ moves towards Q. When ω is tangent to δ_Q then its preimage ω^{-1} is characterized by a cusp point in Q. By moving further ω until it crosses δ_Q in two points, then ω^{-1} forms a loop with a double point at Q.

Now if the arc we are considering is a portion of a basin boundary, we can state the following:

Proposition 2 *(Bischi et al., 1999). When a basin boundary has a contact with a prefocal curve, a particular type of basin bifurcation occurs, causing the creation of cusp points and, after the crossing, of loops, along the basin boundary.*

Let us introduce the first economic example to better understand these results.

2.1.1. The Bischi and Tramontana (2005) model In 2005 Bischi and Tramontana consider the problem of a boundedly rational consumer who adjusts her demand adaptively by updating the quantity of good consumed in the direction indicated by discrepancy the current price and the observed relative utility gain. They also consider the preference of the good not exogenous but endogenously determined by the consumption of the good itself.

The evolution in discrete steps of the quantity (x) consumed and the preference (α) for the good, is regulated by the following map of the plane:

$$T_1 : \begin{cases} x' = F(x,\alpha) = \frac{N(x,\alpha)}{D(x,\alpha)} = \frac{(1-\alpha)x^2+\alpha\mu m - x\mu p}{x(1-\alpha)} \\ \alpha' = G(x) = \frac{1}{k_1+k_2 k_3^x} \end{cases}, \tag{4}$$

where $\mu > 0$ measures how reactive are consumers in adjusting their demand, $m > 0$ is the income of the consumer, $p > 0$ is the exogenous price of the good, and the remaining parameters $k_1 > 1$, $k_2 > 0$ and $0 < k_3 < 1$ permit to shape the relation between consumption and preference.

Map T_1 has in its first component a denominator that vanishes along the lines:

$$x = 0 \quad \text{and} \quad \alpha = 1$$

which are the *set of non-definition* δ_S of the map. In particular, the map takes the form 0/0 in the simple focal point $Q = (0, 0)$, whose prefocal line is

$$\delta_Q = \left\{ (x,\alpha) | \, \alpha = \frac{1}{k_1 + k_2} \right\}.$$

A consequence of Proposition 2 is that in the presence of a focal point the shape of the basins of attraction may be strongly influenced by it. We have an example by considering the following set of parameters: $\mu = 0.5$, $m = 400$, $p = 8$, $k_1 = 17$, $k_2 = 5$ and $k_3 = 0.1$. The basins of attractions of al fixed point corresponding to a rational decision ($\mathcal{F}_r$, in light blue) and of diverging trajectories ($\mathcal{F}_\infty$, in grey) are represented in Fig.1. As we can see the border of $\mathcal{F}_r$ is close to the prefocal curve. In fact, the preimages of the border are close to the focal point. By increasing k_2 to a value of 15 we are at the tangency condition and the cusp in the focal point is created (Fig. 2). After the crossing, with $k_2 = 90$ for instance, it is clearly visible the loop created by the preimages of the portion of the basin boundary that have crossed δ_Q (Fig. 3). Loops like that are called *lobes*.

The shape of the basins of attraction can be even more complicated when a map with vanishing denominator is characterized by a multiplicity of focal points or when basins of different coexisting attractors are intermingled. Two good examples are provided by the next economic models.

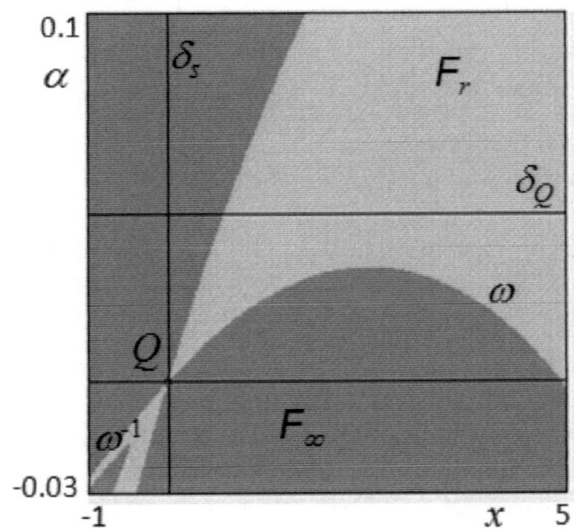

Figure 1. The basin boundary is close to the prefocal line.

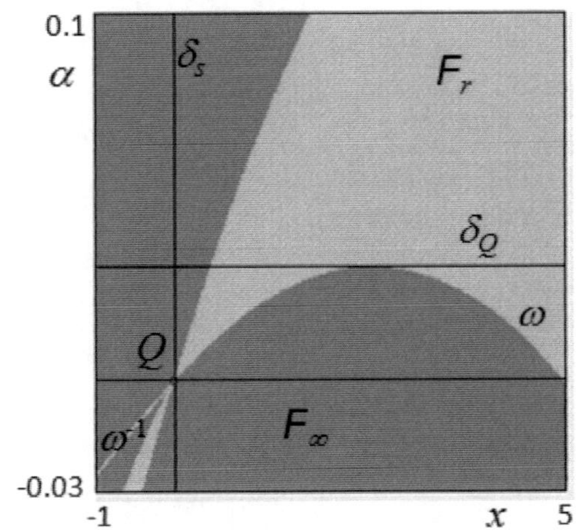

Figure 2. The basin boundary is tangent to the prefocal line.

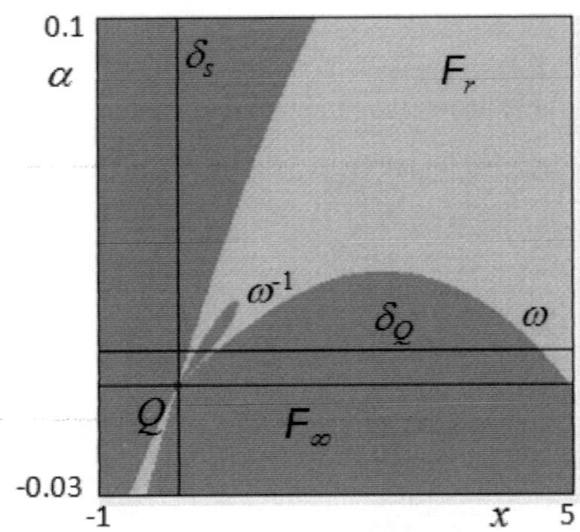

Figure 3. The basin boundary has crossed the prefocal line and a lobe issues from the focal point.

2.1.2. The Bischi and Naimzada (1997) model This model is of particular relevance because it is the one that originated the study of maps with vanishing denominator. Bischi and Naimzada (1997) were studying a model where agents are endowed with a fading memory and try to forecast the next realization of a particular market price by using the information concerning past prices, when they found unusual shapes of the basins of attractions and decided to deepen the mathematical analysis of this map. They realized that the particular basins' structure was a consequence of the presence of a denominator that can vanish in the two-dimensional dynamical system and its dynamic properties were studied in Gardini & Bischi (1996) and Bischi & Gardini (1997). The map regulates the evolution of the expected price (z) and another dynamic variable

Journal of Physics: Conference Series **692** (2016) 012006 doi:10.1088/1742-6596/692/1/012006

(W) useful to model the fading memory of economic agents. This is the map:

$$T_2 : \begin{cases} z' = F(z,W) = \frac{N(z,W)}{D(z,W)} = \frac{\rho W z + \beta z^2 + \delta}{1-\rho W} \\ W' = G(W) = 1 + \rho W \end{cases}, \tag{5}$$

with the restriction: $\rho \in [0,1]$.

Map T_2 in its first component presents a denominator that vanishes in the line of equation:

$$\delta_s : W = -\frac{1}{\rho}.$$

In this case $F(z,W)$ may assume the form $0/0$ in two points:

$$Q_1\left(\frac{1+\sqrt{1-4\beta\delta}}{2\beta}, -\frac{1}{\rho}\right) \text{ and } Q_2\left(\frac{1-\sqrt{1-4\beta\delta}}{2\beta}, -\frac{1}{\rho}\right),$$

both characterized by the same prefocal line:

$$\delta_Q : W = 0,$$

so when an invariant set, such as a basin boundary, crosses the prefocal line, two lobes emerge simultaneously from the focal points, as it is clearly visible in Fig.4 (obtained with $\rho = 0.6$, $\delta = 3$ and $\beta = -1.9$). By reducing the value of β also these lobes cross the prefocal line and other secondary lobes emerge from the focal points and if they cross the prefocal line the structure of the basins may become more complicated, as so the one represented in Fig.5 (where $\beta = -4.1$).

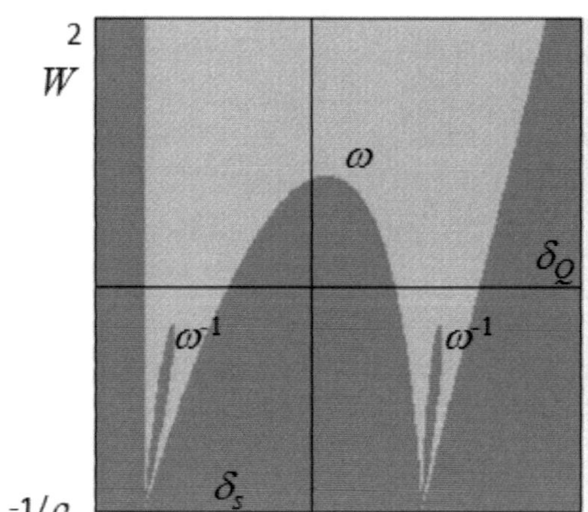

Figure 4. The light blue basin denotes convergence to a fixed point. Starting from the grey region trajectories diverge.

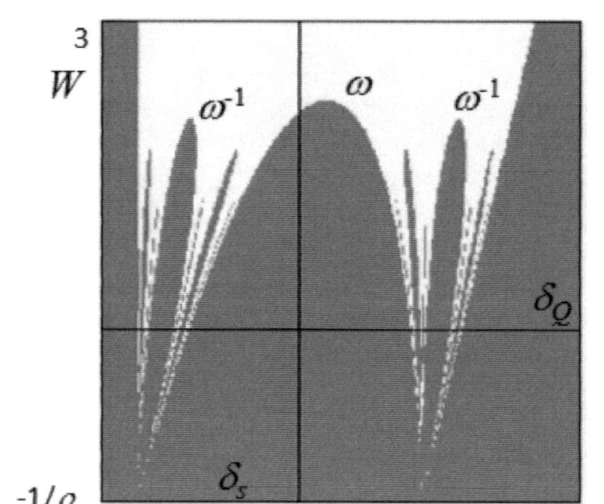

Figure 5. The yellow basin denotes convergence to a cycle of period 3. Starting from the grey region trajectories diverge.

2.1.3. The Naimzada and Tramontana (2009) model Naimzada and Tramontana (2009) study a model of boundedly rational consumer that is a variation of Bischi and Tramontana (2005) that we have already seen. In particular they introduce a different decisional mechanism characterized

Journal of Physics: Conference Series **692** (2016) 012006 doi:10.1088/1742-6596/692/1/012006

by an adjustment towards the direction indicated by the marginal utility currently achieved by the consumer. The dynamical system in the consumption (x) and preference (α) is the following:

$$T_3 : \begin{cases} x' = F(x,y) = x + \gamma \left[\frac{N(x,y)}{D(x,y)}\right] = x + \gamma \left[\frac{\alpha(m-px)x^{1-2\alpha} - p(1-\alpha)x}{x^{1-\alpha}(m-px)^{\alpha}}\right] \\ \alpha' = G(x) = \frac{1}{k_1 + k_2 k_3^x} \end{cases}, \tag{6}$$

where $\gamma > 0$ is the speed of adjustment and the meaning of the other parameters is the same of Bischi and Tramontana (2005). The set of non-definition of this map with denominator is given by:

$$\delta_S : x = 0 \cup x = \frac{m}{p},$$

and we can identify two focal points:

$$Q_1\,(0,0) \quad \text{and} \quad Q_2\left(\tfrac{m}{p}, 1\right),$$

whose prefocal lines are respectively:

$$\delta_{Q_1} : \alpha = \frac{1}{k_1 + k_2} \quad \text{and} \quad \delta_{Q_2} : \alpha = \frac{1}{k_1 + k_2 k_3^{m/p}}.$$

The presence of two focal points with different prefocal lines associated with the coexistence of several locally stable attractors for the map T_3 may give rise to quite complicated basins of attraction like those represented in Fig.s 6 and 7, obtained by keeping fixed $m = 18$, $\gamma = 0.9$ and $k_3 = 0.307$.

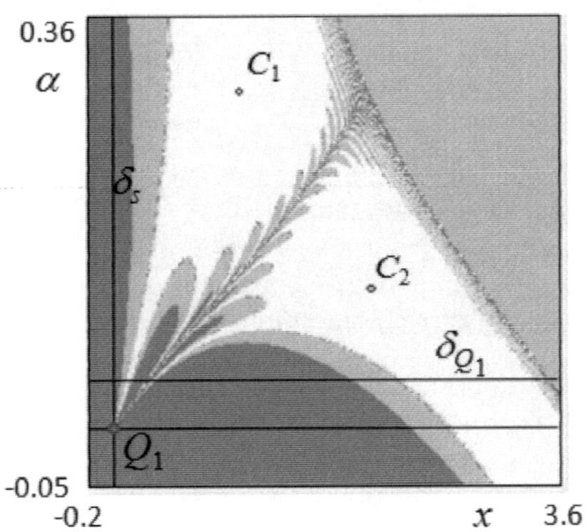

Figure 6. The light blue basin denotes convergence to the fixed point E, while the yellow one convergence to a cycle of period 2 (C_1 and C_2). Starting from the grey region trajectories diverge. We used $p = 3.525$, $k_1 = 1.5$ and $k_2 = 22.555$.

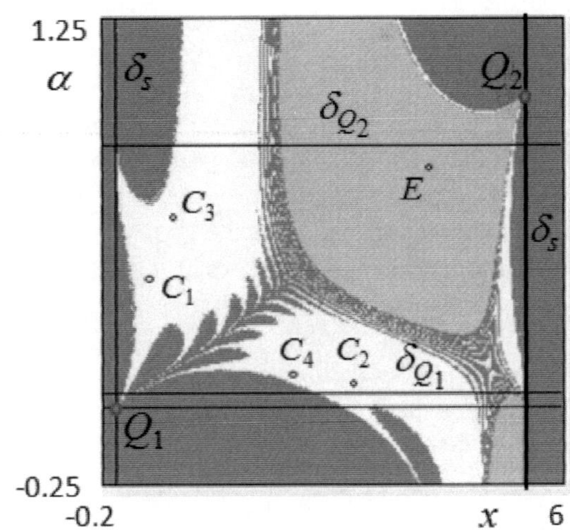

Figure 7. The yellow basin here denotes convergence to a cycle of period 4. We used $p = 3.29$, $k_1 = 1.157$ and $k_2 = 20.4$.

Journal of Physics: Conference Series **692** (2016) 012006 doi:10.1088/1742-6596/692/1/012006

2.1.4. The Foroni et al. (2003) model Three focal points characterize the map studied by Foroni et al. (2003). They consider a cobweb market for the price of a perishable-good. The authors focus on the case of fish but it can be adapted for any other renewable resource market. The producers make their choices on the basis of the expected price of the fish, with a fading memory procedure similar to the one already seen in the Bischi and Naimzada (1997) model (see sec. 2.1.2). By imposing the market clearing conditions they obtain a two-dimensional system regulating the dynamics of the expected price (z) and the memory variable (W):

$$T_4 : \begin{cases} z' = F(z, W) = \frac{N(z,W)}{D(z,W)} = \frac{\rho W z + H(z)}{1-\rho W} \\ W' = G(W) = 1 + \rho W \end{cases}, \tag{7}$$

where $\rho \in [0, 1]$ regulates how fast the memory decays. The set of nondefinition is given by:

$$\delta_s : W = -\frac{1}{\rho}$$

and a focal point of the map T_4 must necessarily verify the condition:

$$H(z) = z \tag{8}$$

Foroni et al. select the following specification of the function $H(z)$:

$$H(z) = \frac{A - S(z)}{B},$$

with

$$S(z) = rg(z)\left(1 - \frac{g(z)}{k}\right)$$

and

$$g(z) = \frac{k}{4}\left\{1 + \frac{c}{qkz} - \frac{\delta}{r} + \sqrt{\left(1 + \frac{c}{qkz} - \frac{\delta}{r}\right)^2 + \frac{8c\delta}{qkrz}}\right\}$$

where $S(z)$ derives from the assumption of logistic intrinsic growth rate of the fish and $g(z)$ comes from the result of an optimization problem for the fisher. We refer the reader to Foroni et al. (2003) for a complete explanation of the meaning of the parameters.

For our pourposes it is important to specifiy that the particular specification of $H(z)$ leads to three roots of the equation (8), that is to three focal points whose coordinates can only be found numerically. The prefocal line is the same for all the three focal points, given by:

$$\delta_Q : W = 0.$$

It is possibile to find a combination of parameters such that two attractors coexist: a fixed point and a periodic cycle. The structure of their basins of attraction is highly influenced by the presence of three focal points. An example is provided in Fig. 8.

2.2. Main results and applications when the focal point is not simple

By using the definition of simple focal point (3) we can easily define the condition that must be verified in order to have a *non simple focal point*:

$$\overline{N}_x \overline{D}_y - \overline{N}_y \overline{D}_x = 0. \tag{9}$$

Bischi et al. (2005) distinguish among various subcases in which (9) holds, considering the number of zeroes in the matrix of first-order partial derivatives of the map.

A subcase that is particularly important for the shape of invariant sets crossing prefocal curves is verified whenever all the first-order partial derivatives vanish in the focal point. In this case we have the next result:

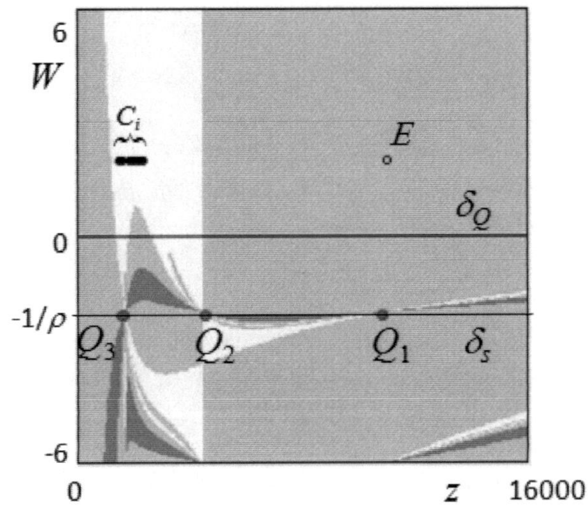

Figure 8. The light blue basin correspondes to convergence to the fixed point E, while the yellow basin is associated to a periodic cycle whose points are labeled C_i. Starting from the grey region trajectories diverge. We used $A = 5241$, $B = 0.28$, $c = 5000$, $k = 400000$, $q = 0.000014$, $r = 0.05$, $\rho = 0.5$ and $\delta = 0.12$.

Proposition 3 *(Bischi et al., 2005). When a map with vanishing denominator (1) has a non simple focal point Q such that $\overline{N}_x = \overline{D}_x = \overline{N}_y = \overline{D}_y = 0$, then:*

- *the prefocal set δ_Q associated with Q belongs to the line $x = G(Q)$*
- *if the second-order partial derivatives of $D(x, y)$ in Q are not all vanishing, then the correspondence between the slope m of an arc γ through Q and the y-coordinate of the point $(G(Q), y(m))$ is generally two-to-one.*

So we can conclude that strucures such as lobes arise by twos and a corollary of such a proposition is the following:

Corollary. *(Bischi et al. 2001b) If the focal point Q belongs to its prefocal curve δ_Q then any arc transverse to δ_Q has infinitely many preimages which are arcs through Q with slopes 0 and ∞ in Q.*

If we are in the situation where Proposition 3 holds and the focal point belongs to its prefocal curves, the loops issuing from the focal point after the global bifurcation are infinitely many. When this arc crossing the prefocal curve is the border of a basin of attraction the basins' structure can be extremely complicated, as it happens in the next economic application.

2.2.1. The Bischi-Kopel-Naimzada (2001) model Bischi et al. (2001b) analyze a so-called *rent-seeking game*, that is a situation in which some agents (or players) compete for obtaining a prize (such as a patent or a contract with the government). In their paper the contenders are two boundedly rational firms that choose production efforts to get the highest profit. The firms are boundedly rational in the sense that they are not able to solve the maximization problem (either because of a lack of information or a lack of computational skills) and dynamically adjust their quantities by using the current marginal profit as a signal.

The two-dimensional dynamical system determining the quantities of the firms is the following:

$$T_5 : \begin{cases} q_1' = F(q_1, q_2) = q_1 \left[1 - c_1 v_1 + v_1 \frac{q_2}{(q_1+q_2)^2}\right] \\ q_2' = G(q_1, q_2) = q_2 \left[1 - c_2 v_2 + v_2 \frac{q_1}{(q_1+q_2)^2}\right] \end{cases}, \tag{10}$$

where the positive parameters c_i $(i = 1, 2)$ represent the marginal costs of the two firms and $v_i (i = 1, 2)$ the speeds of adjustment.

Map T_5 has both the components with a denominator that vanishes along the line:

$$\delta_s : q_2 = -q_1.$$

The map assumes the form 0/0 in correspondence of the focal point $Q(0, 0)$. Bischi et al. (2001) proved that the prefocal set δ_Q is the half-line:

$$\delta_Q : q_2 = \frac{v_2}{v_1} q_1 \text{ where } q_1 \leq \frac{v_1}{4} \text{ and endpoint given by } C\left(\frac{v_1}{4}, \frac{v_2}{4}\right).$$

The focal point Q is not simple and in particular all the first-order partial derivates calculated in Q are equal to zero. Moreover it is easy to see that Q belongs to δ_Q. So we are exactly in the case described by Proposition 3 and in the realm of application of its corollary. We expect to see infinitely many lobes whenever the boundary of a basin of attraction crosses the pre-focal line. In fact, this is what we can see in Fig.9 where it is possible to see the so-called *crescents* (i.e. lobes merged with other lobes) and in its enlargment in Fig. 10 where new lobes are visible between other lobes. We know that they are infinitely many.

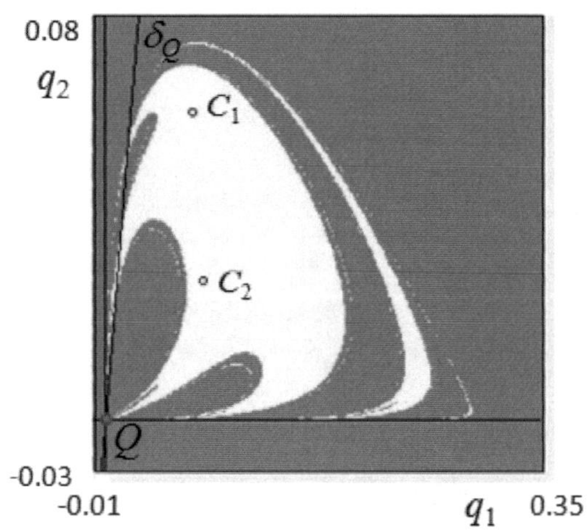

Figure 9. The yellow basin denotes convergence to a cycle of period 2. Starting from the grey region trajectories diverge. We used $c_1 = 3$, $c_2 = 5$, $v_1 = 0.2$ and $v_2 = 0.6$.

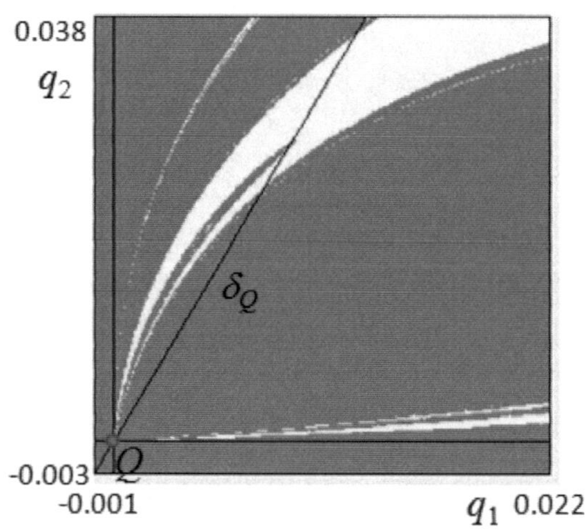

Figure 10. Enlargment of the region around the focal point Q.

2.3. The role of noninvertibility

Until now we have not taken into consideration the role played by the noninvertibility of a map on the structure of invariant sets and basins of attraction. We know that if a map T is noninvertible, then some points of the phase plane (or more generally phase space) are characterized by a number of rank-1 preimages higher than one. This feature often determines a strcture of the basins of attraction that can be quite complicated (white *holes,* or non connected portions, for instance), and if it is combined with the presence of a denominator that can vanish the results can be even more peculiar. The last example, which is also the most recent, belongs to this case.

2.3.1. The Cavalli et al. (2015) model Cavalli et al. (2015) recently proposed a two-dimensional map that describes the evolution of the quantities produced by two boundedly rational and heterogeneous duopolists. This is one of the first attempts to study an oligopoly model with different kinds of decisional mechanisms, without considering any rational player. One firm adopts a gradient-like mechanism of adjustment similar to those already seen in previous examples, while the second adopts a Local Monopolistic Approximation (i.e. it behaves as if it were a monopolist facing a linear demand function, see Bischi et al. (2007)).

The map they obtain is the following:

$$T_6 : \begin{cases} q_1' = F(q_1, q_2) = q_1 + \alpha q_1 \left[\frac{q_2}{(q_1+q_2)^2} - c_1 \right] \\ q_2' = G(q_1, q_2) = \frac{1}{2} q_2 + \frac{1}{2}(q_1 + q_2)\left[1 - c_2 (q_1 + q_2)\right] \end{cases}, \tag{11}$$

where $c_i > 0$ $(i = 1, 2)$ are the marginal costs and $\alpha > 0$ the speed of adjustment of the first duopolist.

The denominator of the first difference equation vanishes along the line:

$$\delta_s : q_2 = -q_1$$

and there exists a focal point $Q(0, 0)$ belonging to its prefocal line, given by:

$$\delta_Q : q_2 = 0.$$

We are in the scenario described in the subsection (2.2), so whenever a lobe appears, infinitely many appear as well. Moreover the focal point may be located in a region of the phase plane whose points have many rank-1 preimages, leading to a repetition of the structure that emerges around the focal point. This is what occurs in Fig.11, where the focal point is in a region whose points have four preimages. Two of the rank-1 preimages (Q_a^{-1} and Q_b^{-1}) are the vertex of the pseudo-triangle where the structure of lobes characterizing the focal point is replicated.

3. Conclusions

In this work we have surveyed some economic applications where plane maps with vanishing denominator regulate the dynamics of the relevant variables. This permitted us on the one hand to summarize the main results obtained through the deep study of this class of maps, and on the other hand to emphasize how important can be for the application to global bifurcations related to the appearence of strucutres related to the so-called focal points of the map. We omitted, for lack of space, to mention the papers such as those of Billings & Curry (1996), Gardini et al. (1999), Gu & Ruan (2005) and Fischer & Gillis (2006), that, dealing with problems like finding the roots of a rational function, have important implications for applications, including those in economics. We hope our survey will be useful to those researchers that build a dynamical system characterized by a map with denominator and want to deepen the global consequences of such a feature.

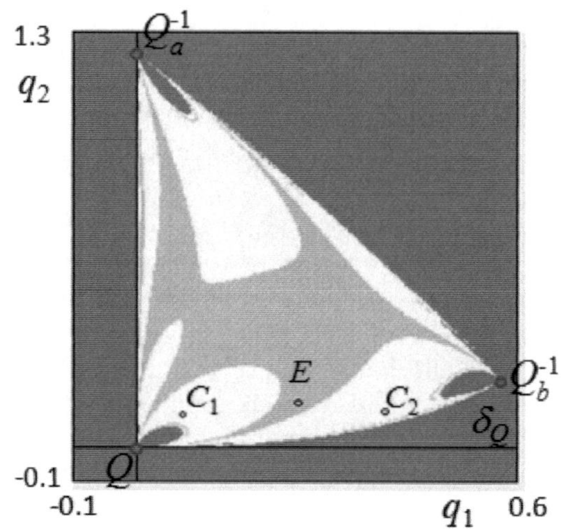

Figure 11. The light blue basin correspondes to convergence to the fixed point E, while the yellow basin is associated to a cycle of period 2. Starting from the grey region trajectories diverge. We used $c_1 = 0.9$, $c_2 = 1.62$ and $\alpha = 1.8$.

References

[1] Billings L and Curry JH 1996 *Chaos* **6** 108-119
[2] Bischi GI and Gardini L 1997 *Int. J. Bif. Chaos* **7** 1555-1577
[3] Bischi GI, Gardini L and Mira C 1999 *Int. J. Bif. Chaos* **9** 119-153
[4] Bischi GI, Gardini L and Mira C 2001a *Nonlinear Anal.* **47** 2171-2185
[5] Bischi GI, Gardini L and Mira C 2003 *Int. J. Bif. Chaos* **13** 2253-2277
[6] Bischi GI, Gardini L and Mira C 2005 *Int. J. Bif. Chaos* **15** 451-496
[7] Bischi GI, Kopel M and Naimzada A 2001b *Nonlinear Anal.* **47** 5309-5324
[8] Bischi GI and Naimzada A 1997 *Econ. Notes* **26** 445-476
[9] Bischi GI, Naimzada A, Sbragia L 2007 *J. Econ. Behav. Organ.* **62** 371-388
[10] Bischi GI and Tramontana F 2005 *Pure Math. Applications* **16** 345-363
[11] Cavalli F, Naimzada A and Tramontana F 2015 *Commun. Nonlinear Science Numer. Simul.* **18** 245-262
[12] Djellit I, Sahari ML and Hachemi A 2013 *J. Applied Anal. Comput.* **3** 11-20
[13] Fischer P and Gillis D 2006 *Int. J. Bif. Chaos* **16** 179-183
[14] Foroni I, Gardini L and Barkley Rosser Jr. J 2003 *Math. Comput. Simul.* **63** 541-567
[15] Gardini L and Bischi GI 1996 *Grazer Mathematische Berichte* 225-231
[16] Gardini L, Bischi GI and Fournier-Prunaret D 1999 *Chaos* **9** 367-380
[17] Gardini L, Bischi GI and Mira C 2007 *Scholarpedia* **2**(9):3277
[18] Gu EG 2009 *Commun. Nonlinear Science Numer. Simul.* **14** 2720-2727
[19] Gu EG and Hao YD 2007 *Chaos Solitons Fractals* **34** 1272-1284
[20] Gu EG and Huang J 2006 *Int. J. Bif. Chaos* **16** 2601-2613
[21] Gu EG and Ruan J 2005 *Int. J. Bif. Chaos* **15** 1625-1639
[22] Mira C 1981 *Comptes Rendus Acad. Sci. Paris Serie I* **292** 146-151
[23] Mira C 1996 *Proceedings of European Conference on iteration theory,* ECIT96, Urbino (8-14 Sept. 1996)
[24] Naimzada A and Tramontana F 2009 *Int. J. Bif. Chaos* **19** 2059-2071

Von Bertalanffy's dynamics under a polynomial correction: Allee effect and big bang bifurcation

J. Leonel Rocha[1], A.K. Taha[2] and D. Fournier-Prunaret[3]

[1] ISEL-Instituto Superior de Engenharia de Lisboa, ADM, Rua Conselheiro Emídio Navarro 1, 1959-007 Lisboa, Portugal
[2] INSA, University of Toulouse, 135 Avenue du Rangueil, 31077 Toulouse, France
[3] LAAS-CNRS, INSA, University of Toulouse, 7 Avenue du Colonel Roche, 31077 Toulouse, France

E-mail: jrocha@adm.isel.pt

Abstract. In this work we consider new one-dimensional populational discrete dynamical systems in which the growth of the population is described by a family of von Bertalanffy's functions, as a dynamical approach to von Bertalanffy's growth equation. The purpose of introducing Allee effect in those models is satisfied under a correction factor of polynomial type. We study classes of von Bertalanffy's functions with different types of Allee effect: strong and weak Allee's functions. Dependent on the variation of four parameters, von Bertalanffy's functions also includes another class of important functions: functions with no Allee effect. The complex bifurcation structures of these von Bertalanffy's functions is investigated in detail. We verified that this family of functions has particular bifurcation structures: the big bang bifurcation of the so-called "box-within-a-box" type. The big bang bifurcation is associated to the asymptotic weight or carrying capacity. This work is a contribution to the study of the big bang bifurcation analysis for continuous maps and their relationship with explosion birth and extinction phenomena.

1. Introduction and motivation

Von Bertalanffy's equation remains one of the most popular flexible growth equations to describe the growth of marine populations, in particular fishes, seabirds, marine mammals, invertebrates, reptiles and sea turtles. The growth equation introduced by von Bertalanffy to model fish weight growth is given by the differential equation

$$\frac{dW_t}{dt} = \frac{V}{3} W_t^{\frac{2}{3}} \left(1 - \left(\frac{W_t}{W_\infty}\right)^{\frac{1}{3}}\right), \tag{1}$$

where W_t is the weight at age t, W_∞ is the asymptotic weight or carrying capacity, $V > 0$ is von Bertalanffy's growth rate constant and t_0 is the theoretical age the chick would have at weight zero, see [5].

Currently the extinction of certain marine species is one of the most worrying problems in ecological and biological research. The Allee effect is an important dynamic phenomenon first described by Allee in 1931, [1]. A demographic Allee effect occurs when individual fitness, at low densities, increases with population density. However, the von Bertalanffy growth model, Eq.(1), do not exhibit the Allee effect, because the associated *per capita* growth rate decreases at low densities. This drawback was corrected using suitable corrections in the work presented by Rocha *et al* in [8]. One of the corrected von Bertalanffy's model proposed is defined by the differential equation,

$$\frac{dW_t}{dt} = \frac{V}{3} W_t^{\frac{2}{3}} \left(1 - \left(\frac{W_t}{W_\infty}\right)^{\frac{1}{3}}\right) \frac{W_t - E}{W_\infty + C}, \tag{2}$$

with $|E| < W_\infty$ and $C > 0$, where E is the rarefaction critical density (density threshold) or Allee limit and W_∞ is the carrying capacity. Note that the adjustment or correction factor is of polynomial type.

2. von Bertalanffy's functions of polynomial type

In the present work, we study a dynamical approach to von Bertalanffy's growth models: a new class of functions describing the existence or not of strong and weak Allee effect is investigated. Consider the family of functions $f : [0,1] \to]-\infty, 1]$, defined by

$$f(x; r, K, E, C) = r\, x^{\frac{2}{3}} \left(1 - x^{\frac{1}{3}}\right) \frac{Kx - E}{K + C}, \tag{3}$$

with $x = \frac{W_t}{W_\infty} = \frac{W_t}{K} \in [0,1]$ the normalized weight, $r = \frac{V}{3} \times K^{\frac{2}{3}} > 0$ an intrinsic growth rate of the individual weight, the Allee limit $E \in \mathbf{R}$ and the carrying capacity $K > 0$ satisfying $|E| < K$ and $C > -K$. This family of functions will be designated by von Bertalanffy's functions of polynomial type, see some examples in Figs.1, 2 and 3. Generically, these functions are unimodal or bimodal maps for which extinction is inevitable for too high or too low initial population densities. Furthermore, this family of maps is proportional to the right hand side of the von Bertalanffy's growth model, Eq.(2).

Let $A_r = f(A_r; r, K, E, C)$ be a fixed point of f and $A_r^* = max\{f^{-1}(A_r; r, K, E, C)\}$. From Eq.(3), we have that $E/K < A_r$. The interval $[0, E/K]$ will be designated by extinction interval and E/K is the Allee threshold. The following properties are satisfied:

(A1) f is continuous on $[0,1]$;

(A2) there exists $[A_r, A_r^*] \subset [0,1] : f(x; r, K, E, C) \leq A_r, \forall x \in [0,1] \setminus]A_r, A_r^*[$;

(A3) f has an unique critical point $c \in]A_r, A_r^*[$;

(A4) $f \in C^3(]A_r, A_r^*[)$;

(A5) $f'(x; r, K, E, C) \neq 0, \forall x \in]A_r, A_r^*[\setminus \{c\}$, $f'(c; r, K, E, C) = 0$ and $f''(c; r, K, E, C) < 0$;

(A6) the Schwarzian derivative $S(f(x; r, K, E, C)) = \frac{f'''(x;r,K,E,C)}{f'(x;r,K,E,C)} - \frac{3}{2}\left(\frac{f''(x;r,K,E,C)}{f'(x;r,K,E,C)}\right)^2$ verifies the following properties:

(i) $S(f(x; r, K, E, C)) < 0, \forall x \in]A_r, A_r^*[\setminus \{c\}$ and $S(f(c; r, K, E, C)) = -\infty$;
(ii) $S(f(x \to 0^+; r, K, E, C)) > 0$, with $|E| < K \wedge E \neq 0$;
(iii) $S(f(x \to 0^+; r, K, E, C)) < 0$, with $E = 0$.

To von Bertalanffy's functions, Eq.(3), we define a class of functions that describes the growth of a population with Allee effect.

Definition 1: Consider $0 \leq E < K$, $C > -K$, $r > 0$ and $f(x; r, K, E, C)$ defined by Eq.(3). von Bertalanffy's functions $f : [0,1] \to]-\infty, 1]$ attaining its maximum at c are Allee's functions, if there are real numbers A_r and B_r such that:

(i) $0 < A_r < c < B_r < 1$;

(ii) $\forall x \in [0,1] \setminus [A_r, B_r] \Rightarrow f(x; r, K, E, C) < x$;

(iii) $\forall x \in]A_r, B_r[\Rightarrow f(x; r, K, E, C) > x$.

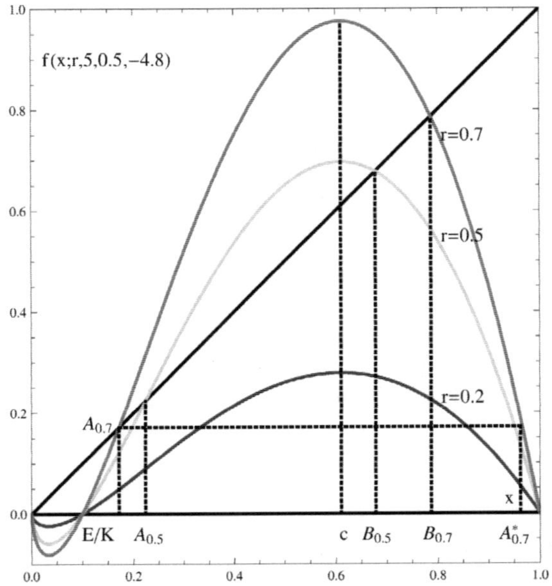

Figure 1. von Bertalanffy's functions with strong Allee effect $f(x; r, K, E, C)$ at $K = 5$, $E = 0.5$, $C = -4.8$ and $r = 0.2$ (1 fix. pt.: 0), $r = 0.5$ (3 fix. pts.: 0, $A_{0.5}$ and $B_{0.5}$) and $r = 0.7$ (3 fix. pts.: 0, $A_{0.7}$ and $B_{0.7}$).

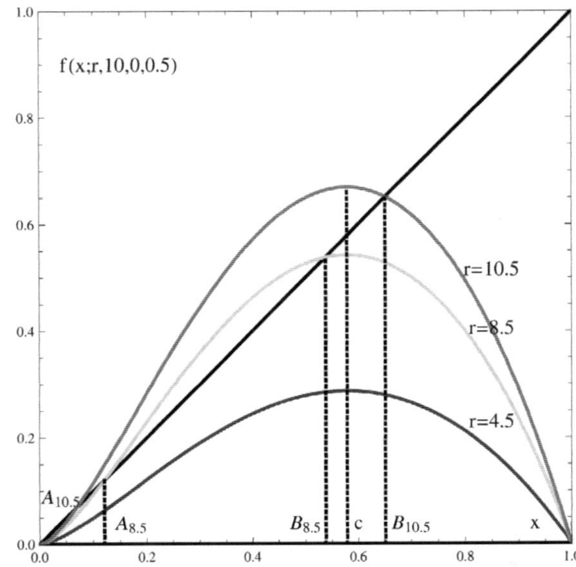

Figure 2. von Bertalanffy's functions with weak Allee effect $f(x; r, K, E, C)$ at $K = 10$, $E = 0$, $C = 0.5$ and $r = 4.5$ (1 fix. pt.: 0), $r = 8.5$ (3 fix. pts.: 0, $A_{8.5}$ and $B_{8.5}$) and $r = 10.5$ (3 fix. pts.: 0, $A_{10.5}$ and $B_{10.5}$).

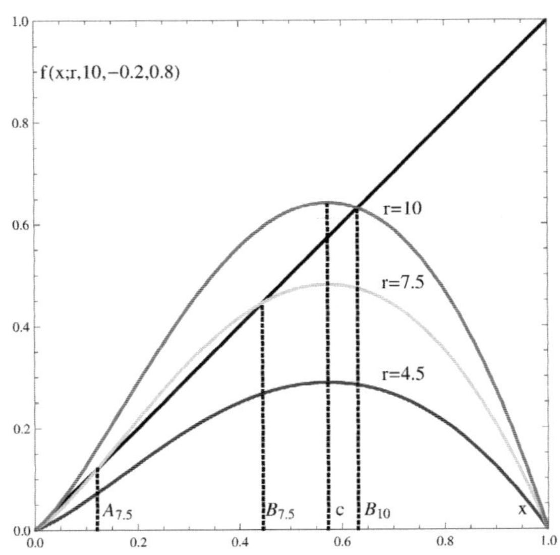

Figure 3. von Bertalanffy's functions with no Allee effect $f(x; r, K, E, C)$ at $K = 10$, $E = -0.2$, $C = 0.8$ and $r = 4.5$ (2 fix. pts.: 0 and O_r), $r = 7.5$ and $r = 10$ (4 fix. pts.: $0, O_r, A_r$ and B_r).

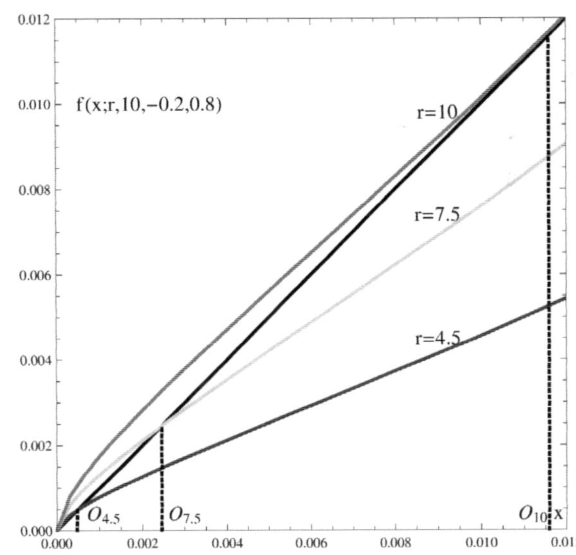

Figure 4. von Bertalanffy's functions with no Allee effect $f(x; r, K, E, C)$ at $K = 10$, $E = -0.2$, $C = 0.8$ and $r = 4.5$ (2 fix. pts.: 0 and O_r), $r = 7.5$ and $r = 10$ (4 fix. pts.: $0, O_r, A_r$ and B_r).

This class of Allee's functions is divided into two subclasses of functions that verify the following properties.

Property 1: Under the conditions of Def.1 it is established that:

(i) If $0 < E < K$ and $C > -K$, the von Bertalanffy functions $f : [0, 1] \to]-\infty, 1]$ are a family

Journal of Physics: Conference Series **692** (2016) 012007 doi:10.1088/1742-6596/692/1/012007

of strong Allee's functions, see Fig.1;

(ii) If $E = 0$ and $C > -K$, the von Bertalanffy functions $f : [0,1] \to [0,1]$ are a family of weak Allee's functions, see Fig.2.

This classification is based on the concepts of strong and weak Allee's effects to *population growth rates*. The Allee effect is described as *strong* when there is a density threshold, where the population growth rate becomes null and below which are negative. The Allee effect is described as *weak* when the populations do not exhibit any thresholds, the point $x = 0$ becomes stable and the population growth rate are positive. von Bertalanffy's functions f also includes another class of important functions, for which the existence of Allee effect has no evidence.

Definition 2: Consider $-K < E < 0$, $C > -K$, $r > 0$ and $f(x; r, K, E, C)$ defined by Eq.(3). von Bertalanffy's functions $f : [0,1] \to [0,1]$ attaining its maximum at c are functions with no Allee effect, if there are real numbers O_r, A_r and B_r such that:

(i) $0 < O_r < A_r < c < B_r < 1$;

(ii) $\forall x \in (]O_r, A_r[\cup]B_r, 1]) \Rightarrow f(x; r, K, E, C) < x$;

(iii) $\forall x \in (]0, O_r[\cup]A_r, B_r[) \Rightarrow f(x; r, K, E, C) > x$.

From Def.2, the fixed point $x = 0$ is always unstable. It is for this reason that this class of functions with no Allee effect does not contemplate the phenomenon of extinction and any thresholds. Note that from Def.1 and Prop.1 it follows that, in the cases of strong and weak Allee's functions, we will have at maximum three fixed points $0, A_r$ and B_r, see Figs.1 and 2. On the other hand, from Def.2, in the case of functions with no Allee effect, we will have at least two fixed points 0 and O_r, and at most four fixed points $0, O_r, A_r$ and B_r, depending on the evolution of the parameter r, see Fig.3, for details of the fixed point O_r see Fig.4.

Remark 1: Note that the transition from strong Allee effect ($0 < E < K$ and $C > -K$) to no Allee effect ($-K < E < 0$ and $C > -K$), passing through the weak Allee effect ($E = 0$ and $C > -K$), involve the several parameters considered in the definition of the new models, standing out the Allee limit E and the parameter C. See in Table 1 the summary of the most important properties for von Bertalanffy's functions of polynomial type.

Remark 2: Note that for von Bertalanffy's functions f when r varies monotonically in the interval $]0, \check{r}[$, where $\check{r}$ is such that $f(c; \check{r}, K, E, C) = 1$, there exists a fixed point B_r such that its multiplier ($s = \frac{\partial f}{\partial x}(x; r, K, E, C)$ calculated at the fixed point) decreases monotonically from $+1$, and a fixed point with s which increases monotonically from $+1$.

3. Bifurcation structures in von Bertalanffy's functions: big bang bifurcation

In this work we investigate in detail the bifurcation structures of von Bertalanffy's functions given by Eq.(3), in the (C, r) two-dimensional parameter space.

3.1. Fold and flip bifurcation curves

We will make use of the classical fold and flip bifurcations. We study the behavior of such curves related with some cycles of order $n \in \mathbf{N}$, see Figs.5, 7 and 9. In a general way, to von Bertalanffy's functions f, the fold and flip bifurcation curves relative to a cycle of order n are determined as follows. If $x \in [0, 1[$ is a point of an order n cycle that satisfies the equations

$$f^n(x; r, K, E, C) = x \quad \text{and} \quad \frac{\partial f^n}{\partial x}(x; r, K, E, C) = 1 \qquad (4)$$

then there exists a solution φ_n, such that the fold bifurcation curves relative to a cycle of order $n \in \mathbf{N}$ are given by $r(C) = \varphi_n(x; K, E, C)$, and are denoted by $\Lambda^j_{(n)_0}$, where j is the number of the curve, which differentiates cycles of the same order.

On the other hand, if $x \in [0,1[$ is such that,

$$f^n(x;r,K,E,C) = x \quad \text{and} \quad \frac{\partial f^n}{\partial x}(x;r,K,E,C) = -1 \tag{5}$$

then exists a solution ψ_n, such that the flip bifurcation curves relative to a cycle of order $n \in \mathbf{N}$ are given by $r(C) = \psi_n(x;K,E,C)$, and are denoted by Λ_n^j, where j is the number of the curve, which differentiates cycles of the same order.

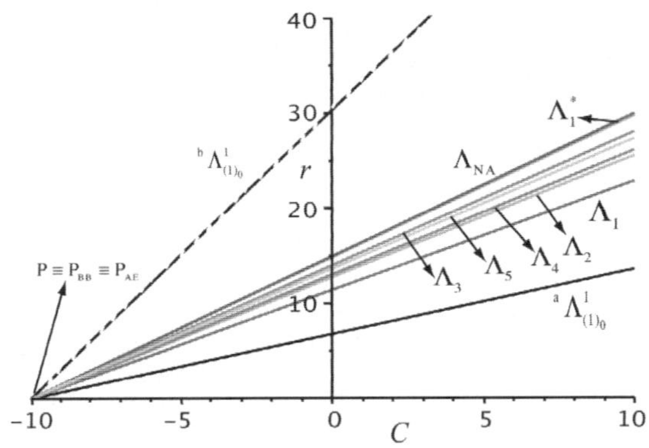

Figure 5. Bifurcations curves of $f^n(x;r,K,E,C)$ with strong Allee effect, for $n = 1,2,3,4,5$, in the (C,r) parameter plane at $K = 10$ and $E = 0.02$. $^a\Lambda^1_{(1)_0}$ and $^b\Lambda^1_{(1)_0}$ are the fold bifurcation curves of different cycles of order $n = 1$, Λ_1, Λ_2, Λ_3, Λ_4 and Λ_5 are the flip bifurcation curves of the cycles of order $n = 1,2,3,4,5$, respectively; Λ_1^* is the chaotic semistability curve and Λ_{NA} is the bifurcation curve of non admissibility, in this case $\Lambda_1^* \neq \Lambda_{NA}$. P_{BB} is the big bang bifurcation point and P_{AE} is the Allee bifurcation point.

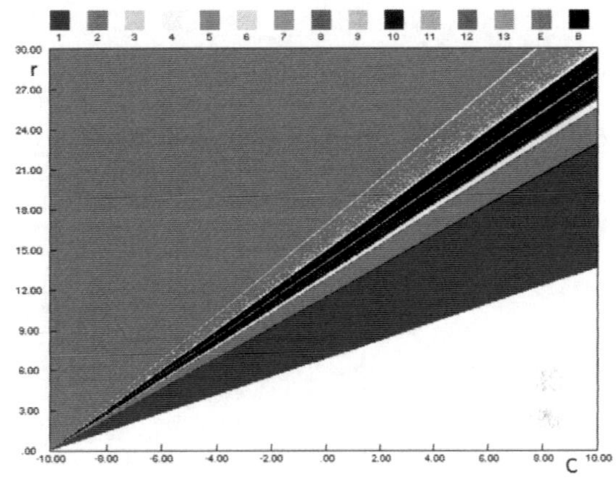

Figure 6. Bifurcation diagram of $f(x;r,K,E,C)$ with strong Allee effect, at $K = 10$ and $E = 0.02$, in the (C,r) parameter plane. The white region is the unconditional extinction region, the blue region is the stability region of the fixed point A_r. The region between the blue and the gray ones corresponds to period doubling region and chaotic region (existence region of cycles as shown on top of figure). The gray region is the no admissible region.

Remark 3: There follows some fundamental properties of the fold and flip bifurcation curves, which allow a better understanding of the bifurcation diagrams of von Bertalanffy's functions, at the (C,r) parameter plane:

(i) in the cases of strong and weak Allee's functions, the fold bifurcation curve $\Lambda_{(1)_0}$ of the fixed points A_r and B_r, for $n = 1$, is the bifurcation curve which defines the transition between the unconditional extinction region and the stability region of the fixed point B_r;

(ii) in the case of functions with no Allee effect the fold bifurcation curves $^a\Lambda^2_{(1)_0}$ and $^b\Lambda^2_{(1)_0}$ characterize the stability of the fixed points O_r and B_r, and the instability of the fixed points 0 and A_r. Reason why for this class of functions does not exist an unconditional extinction region;

(iii) in all cases, the flip bifurcation curve Λ_1, of the stable fixed point B_r, for $n = 1$, is the bifurcation curve which defines the transition between the stability region and the period doubling region. The upper limit of the period doubling region is defined by the accumulation value of the flip bifurcation curves of the cycle of order 2^n, of the stable fixed points nonzero. This bifurcation curve is denoted by Λ_∞, considering $x \in [0,1[$ a fixed point, we have,

$$\Lambda_\infty = \lim_{n \to \infty} \psi_{2^n}(x;K,E,C);$$

(iv) the chaotic region is bounded below by Λ_∞ and is upper bounded by the chaotic semistability curve, which is then studied. In this region are observed all fold and flip bifurcation curves of the cycle of order different than 2^n, Λ_k, with $k \neq 2^n$, identified and ordered in the "box-within-a-box" bifurcation structure.

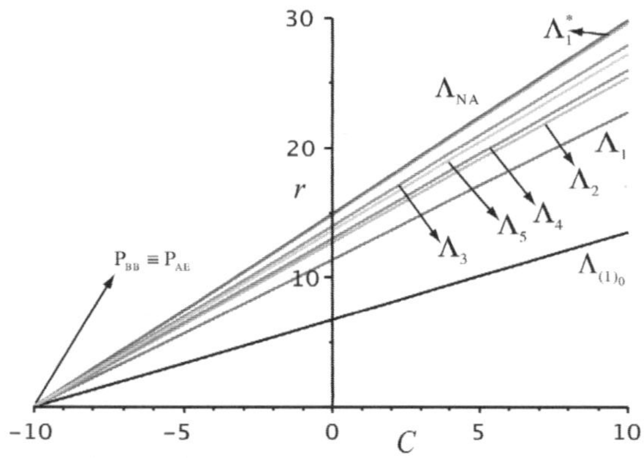

Figure 7. Bifurcations curves of $f^n(x;r,K,E,C)$ with weak Allee effect, for $n=1,2,3,4,5$, in the (C,r) parameter plane at $K=10$ and $E=0$. $\Lambda_{(1)_0}$ is the fold bifurcation curve; Λ_1, Λ_2, Λ_3, Λ_4 and Λ_5 are the flip bifurcation curves of the cycles of order $n=1,2,3,4,5$, respectively; Λ_1^* is the chaotic semistability curve and Λ_{NA} is the bifurcation curve of non admissibility, in this case $\Lambda_1^* \neq \Lambda_{NA}$. P_{BB} is the big bang bifurcation point and P_{AE} is the Allee bifurcation point.

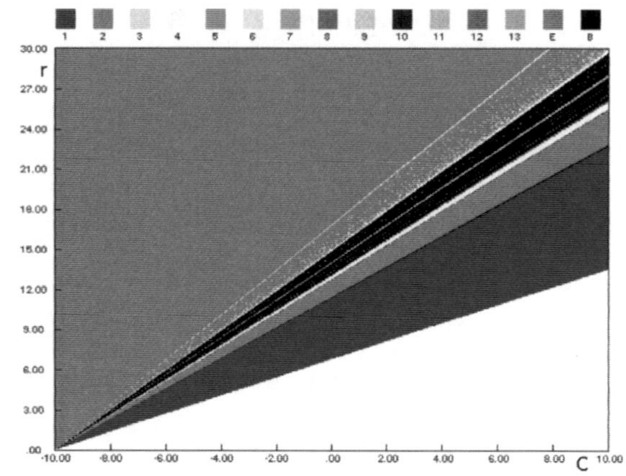

Figure 8. Bifurcation diagram of $f(x;r,K,E,C)$ with weak Allee effect, at $K=10$ and $E=0$, in the (C,r) parameter plane. The white region is the unconditional extinction region, the blue region is the stability region of the fixed point A_r. The region between the blue and the gray ones corresponds to period doubling region and chaotic region (existence region of cycles as shown on top of figure). The gray region is the no admissible region.

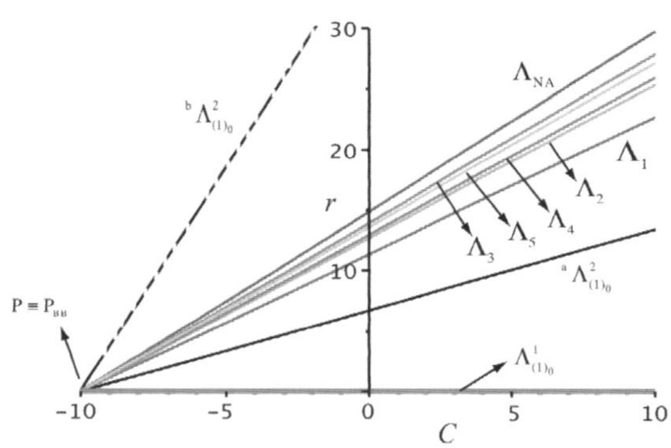

Figure 9. Bifurcations curves of $f^n(x;r,K,E,C)$ with no Allee effect, for $n=1,2,3,4,5$, in the (C,r) parameter plane at $K=10$ and $E=-0.02$. $\Lambda^1_{(1)_0}$, ${}^a\Lambda^2_{(1)_0}$ and ${}^b\Lambda^2_{(1)_0}$ are the fold bifurcation curves of different cycles of order $n=1$; Λ_1, Λ_2, Λ_3, Λ_4 and Λ_5 are the flip bifurcation curves of the cycles of order $n=1,2,3,4,5$, respectively; Λ_{NA} is the bifurcation curve of non admissibility. P_{BB} is the big bang bifurcation point.

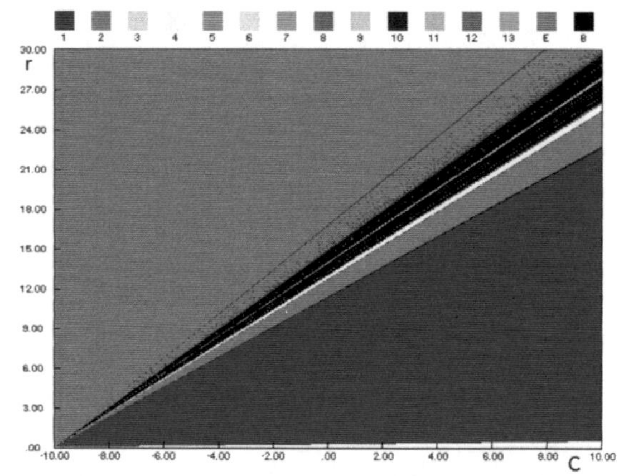

Figure 10. Bifurcation diagram of $f(x;r,K,E,C)$ with no Allee effect, at $K=10$ and $E=-0.02$, in the (C,r) parameter plane. The blue region is the stability region of the fixed point O_r. The region between the blue and the gray ones corresponds to period doubling region and chaotic region (existence region of cycles as shown on top of figure). The gray region is the no admissible region.

3.2. Allee's effect region: chaotic semistability curve and Allee's bifurcation point

The notions of *chaotic semistability curve*, *Allee's effect region* and *Allee's bifurcation point* are fundamental for a detailed characterization of the bifurcation structures of von Bertalanffy's functions.

Definition 3: Let $f(x;r,K,E,C)$ be von Bertalanffy's functions, Eq.(3), with $r>0$, $C>-K$ and $0 \leq E < K$, satisfying $(A1)-(A5)$. The chaotic semistability curve, denoted by

Journal of Physics: Conference Series **692** (2016) 012007 doi:10.1088/1742-6596/692/1/012007

Λ_1^*, is defined by

$$\Lambda_1^* = \left\{(C,r) \in \mathbf{R}^2 : f^2(c;r,K,E,C) = A_r\right\}, \tag{6}$$

where A_r is the first positive fixed point of f and c is the critical point defined in $(A3)$.

This curve is a bifurcation curve for von Bertalanffy's functions with strong and weak Allee effects, see Figs. 5 and 7, because it defines where the chaotic region finishes and begins another region with a different kind of dynamics: the Allee effect region. The Allee effect region is characterized by an essential extinction, i.e., a populational occurrence where the maximum size growth of the population exceeds the critical density and the populations are almost surely doomed to extinction, verifying $f^2(c;r,K,E,C) < A_r$.

Definition 4: Let $f(x;r,K,E,C)$ be von Bertalanffy's functions, Eq.(3), with $r > 0$, $C > -K$ and $0 \leq E < K$, satisfying $(A1) - (A5)$. The Allee effect region, denoted by R_{AE}, is defined by

$$R_{AE} = \left\{(C,r) \in \mathbf{R}^2 : A_r^* < f(c;r,K,E,C) < 1\right\},$$

where $A_r^* = max\{f^{-1}(A_r;r,K,E,C)\}$ and c is the critical point defined in $(A3)$.

To von Bertalanffy's functions with strong and weak Allee effects, the Allee effect region is limited superiorly by the fullshift curve or curve of non admissibility, see Figs. 6 and 8. However, to von Bertalanffy's functions with no Allee effect this curve limit superiorly the chaotic region, see Figs.9 and 10.

Definition 5: Let $f(x;r,K,E,C)$ be von Bertalanffy's functions, Eq.(3), with $r > 0$, $|E| < K$ and $C > -K$, satisfying $(A1) - (A5)$ and under the conditions of Definitions 1 and 2. The non admissibility curve, denoted by Λ_{NA}, is defined by

$$\Lambda_{NA} = \left\{(C,r) \in \mathbf{R}^2 : f(c;r,K,E,C) = 1\right\}, \tag{7}$$

where c is the critical point defined in $(A3)$.

The above definitions allow us to write the following properties.

Property 2: Under the conditions of Definitions 3, 4 and 5 it is established that the existence of Allee's effect region R_{AE} implies that:

(i) if $C > -K$ and $0 \leq E < K$, then Λ_1^* exists and $\Lambda_1^* \neq \Lambda_{NA}$;
(ii) if $C > -K$ and $-K < E < 0$, then Λ_1^* not exists.

Property 3: Under the conditions of Definition 5 it is established that the existence of the non admissibility curve Λ_{NA} implies that:

(i) if $C > -K$ and $0 \leq E < K$, then Λ_{NA} is the bifurcation curve that separates the Allee effect region R_{AE} and the no admissible region;
(ii) if $C > -K$ and $-K < E < 0$, then Λ_{NA} is the bifurcation curve that separates the chaotic region and the no admissible region.

In the cases of strong and weak Allee's functions the bifurcation curves Λ_1^* and Λ_{NA} intersect in a bifurcation point, where it begins the Allee effect region R_{AE}.

Definition 6: Let $f(x;r,K,E,C)$ be von Bertalanffy's functions, Eq.(3), with $r > 0$, $C > -K$ and $0 \leq E < K$, satisfying $(A1) - (A5)$. The Allee bifurcation point, denoted by P_{AE}, is defined by

$$P_{AE} = \left\{(C,r) \in \mathbf{R}^2 : f(c;r,K,E,C) = A_r^* \wedge f(c;r,K,E,C) = 1\right\},$$

where $A_r^* = max\{f^{-1}(A_r;r,K,E,C)\}$ and c is the critical point defined in $(A3)$.

In the (C,r) parameter plane, the Allee bifurcation points are characterized by the symmetric of the carrying capacity K and by a null intrinsic growth rate r, as follows in the next result.

Journal of Physics: Conference Series **692** (2016) 012007 doi:10.1088/1742-6596/692/1/012007

Proposition 1: Let $f(x; r, K, E, C)$ be von Bertalanffy's functions, Eq.(3), with $r > 0$, $C > -K$ and $0 \leq E < K$, satisfying $(A1) - (A5)$. The Allee bifurcation point is given by

$$P_{AE} \equiv (C \to -K^+, r \to 0^+).$$

Proof Consider the intersection of the chaotic semistability curve Λ_1^* with the non admissibility curve Λ_{NA}, given by the following equations:

$$f(A_r; r, K, E, C) = A_r,\ f^2(c; r, K, E, C) = A_r \text{ and } f(c; r, K, E, C) = 1, \tag{8}$$

Consider conditions $(A1) - (A5)$, we verify that the only solution of Eq.(8) is given by $f(1; r, K, E, C) = A_r$. From Eq.(3), we have $f(1; r, K, E, C) = 0 \implies A_r \equiv 0$. So, we can be in a limit case of bifurcation of the fixed point $x = 0$, where two fixed points of f becomes three fixed points. However, by hypothesis this case is not possible, because we are working with von Bertalanffy's functions with strong and weak Allee effects. From Definition 1 and Property 1, we will have at least one fixed point $x = 0$, and at most three fixed points $0, A_r$ and B_r, depending on the evolution of the parameters. Thus, we are in the case of an only fixed point $x = 0$. Consequently, the implicit expression of the fixed points, given by

$$x = \left(r \left(1 - x^{\frac{1}{3}} \right) \frac{Kx - E}{K + C} \right)^3$$

has no meaning, i.e., $C \to -K^+$. Accordingly, so that Definition 1 and Property 1 make sense it is necessary that $r \to 0^+$. Therefore, $P_{AE} \equiv (C \to -K^+, r \to 0^+)$ is the Allee bifurcation point at the (C, r) parameter plane.

3.3. Big bang bifurcation of von Bertalanffy's functions

New kinds of bifurcation points are studied to this family of functions, in the (C, r) parameter plane. These points are called *big bang bifurcation points* and are associated with particular bifurcation structures. Typically, big bang bifurcations occur in the context of piecewise-smooth discontinuous dynamics, whenever two fixed points cross simultaneously the boundary and become virtual. In [2] it was proposed that these specific bifurcations be designated as big bang bifurcations. For more details about big bang bifurcations for discontinuous piecewise maps see for example [3], [4], [6] and references therein. In [9] and [10] are presented the first studies on big bang bifurcation analysis for continuous maps.

As previously indicated in Sec.2, von Bertalanffy's functions f satisfy the properties $(A1) - (A6)$ and Remark 2. In particular, it is also verified that f are continuous maps with respect to the parameter $r > 0$. However, as noted in *(ii)* of $(A6)$, the Schwarzian derivative of f is not negative throughout the interval $[0, 1]$, for some parameter values. This disturbance alters some of the classical properties in the behavior of bifurcations, see for example [9] and [10]. Note that the occurrence *(ii)* of $(A6)$ it is verified in a neighborhood of the fixed point $x = 0$. On the other hand, from Definition 6 and Proposition 1 the bifurcation curves, chaotic semistability curve Λ_1^* and non admissibility curve Λ_{NA}, intersect in the Allee bifurcation point P_{AE} (in a "limit" case of f, with only the fixed point $x = 0$). Nevertheless, from Eqs.(4) and (5), if we consider $n = 1$ and $C \to -K^+$ becomes

$$x = 0,\ \varphi_1 (0; K, E, C \to -K^+) = 0 \text{ and } \psi_1 (0; K, E, C \to -K^+) = 0 \tag{9}$$

This means that the fold bifurcation curve $\Lambda_{(1)_0}$ and the flip bifurcation curve Λ_1 also intersect in the Allee bifurcation point $P_{AE} \equiv (C \to -K^+, r \to 0^+)$. These special characteristics lead us to predict some changes in bifurcation structures of f, restricted to the parameter region

Journal of Physics: Conference Series **692** (2016) 012007 doi:10.1088/1742-6596/692/1/012007

$R = \{(C, r) \in \mathbf{R}^2 : C \to -K^+,\ r > 0\}$, with $C > -K$ and $|E| < K$. Thus, the sufficient conditions required by Mira in [7] for the existence of "box-within-a-box" fractal bifurcations structure are not satisfied for von Bertalanffy's functions, restricted to R.

Remark 4: An interval of existence of an attractive limit set at a finite distance can be defined for f:

(i) if $0 \leq E < K$ and $C > -K$, then this set is defined by

$$\bar{\Omega}_1 = [r_{(1)_0}, r_1^*], \tag{10}$$

where $r_{(1)_0} = \varphi_1(x; K, E, C)$ and $r_1^* = \zeta(C)$ is given by Eq.(6), see Figs.5 and 7;

(ii) if $-K < E < 0$ and $C > -K$, then this set is defined by

$$\Omega_1 = [r_{(1)_0}, r_{NA}], \tag{11}$$

where $r_{(1)_0} = \varphi_1(x; K, E, C)$ and $r_{NA} = \xi(C)$ is given by Eq.(7), see Fig.9.

The sets $\bar{\Omega}_1$ and Ω_1 are called boxes, inside which occur all the possible bifurcations of f, for more details see for example [7].

Property 4: Let $f(x; r, K, E, C)$ be von Bertalanffy's functions, Eq.(3), with $r > 0$ and satisfying $(A1) - (A6)$. If $0 \leq E < K$ and $C > -K$, then $\bar{\Omega}_1 \subset \Omega_1$.

From the above considerations, Eqs.(9), (10) and (11), the bifurcation analysis made from fold and flip bifurcation curves, Figs.5, 7 and 9, and numerical simulations of the bifurcation diagrams, Figs.6,8 and 10, allow us to formalize the following conjecture:

Conjecture 1: Let $f(x; r, K, E, C)$ be von Bertalanffy's functions, Eq.(3), with $r > 0$, $C > -K$ and $|E| < K$, satisfying $(A1) - (A6)$, $\Lambda_{(n)_0}$ be the fold bifurcation curves, given by Eq.(4), and Λ_n be the flip bifurcation curves, given by Eq.(5), with all bifurcation curves relative to a cycle of order $n \in \mathbf{N}$. In the limit case, when $C \to -K^+$, it is verified that the fold bifurcation curves $\Lambda_{(n)_0}$, for $n \geq 1$, and the flip bifurcation curves Λ_n, for $n \geq 1$, intersect on the bifurcation point $P_{BB} = (C \to -K^+, r \to 0^+)$, following the "box-within-a-box" fractal bifurcations structure in the box Ω_1, Eq.(11).

In this context we expose one of the main definition of this work.

Definition 7: In conditions of Conjecture 1, suppose that there is a fold or a flip bifurcation curve of a cycle of order $k \in \mathbf{N}$ that intersect the non admissibility curve Λ_{NA}, then the coordinates of all the cycles of order $j > k$ converge to the coordinates of the cycle of order k. The coordinates of the cycle of order k define a big bang bifurcation point for von Bertalanffy's functions f with the "box-within-a-box" bifurcation structure.

Proposition 2: If Conjecture 1 is satisfied, then the point P_{BB} is a big bang bifurcation point for f. It is verified that:

(i) if $0 \leq E < K$ and $C > -K$, then the fold bifurcation curves $\Lambda_{(n)_0}$, for $n \geq 1$, the flip bifurcation curves Λ_n, for $n \geq 1$ and the chaotic semi-stability curve Λ_1^* are issuing from P_{BB}, following the "box-within-a-box" fractal bifurcations structure in the box $\bar{\Omega}_1$;

(ii) if $-K < E < 0$ and $C > -K$, then the fold bifurcation curves $\Lambda_{(n)_0}$, for $n \geq 1$, the flip bifurcation curves Λ_n, for $n \geq 1$ and the non admissibility curve Λ_{NA} are issuing from P_{BB}, following the "box-within-a-box" fractal bifurcations structure in the box Ω_1.

Proof Consider the limit case of $f(x; r, K, E, C)$, when $C \to -K^+$, i.e., it is verified that $f(x; r \to 0^+, K, E, C \to -K^+) \to 0$, for all initial states in $[0, 1]$. From Eq.(9), $x = 0$ is the only fixed point of f with $r > 0$, $C > -K$ and $|E| < K$. In this limit case, the dynamics cannot escape from $[0, 1]$ (besides a hyperbolic set of zero measure). These conditions are sufficient to state the existence of the big bang bifurcation point $P_{BB} = (C \to -K^+, r \to 0^+)$. Note that the Allee bifurcation points P_{AE} is coincident with the big bang bifurcation point P_{BB}. Therefore, from Conjecture 1, Definitions 6 and 7, Proposition 1 and Property 4 the desired results follows.

Journal of Physics: Conference Series **692** (2016) 012007 doi:10.1088/1742-6596/692/1/012007

4. Conclusion

The main goal of this work has been the definition and investigation of new ones population dynamic discrete models with the growth of the population given by a family of von Bertalanffy's functions with a polynomial correction, from the point of view of stability analysis and bifurcation theory. The Allee bifurcation points and the big bang bifurcation points are associated to the asymptotic weight or carrying capacity of these models. The definitions of chaotic semistability and non admissibility bifurcation curves were crucial in the description of Allee's effect region and Allee's bifurcation points. Table 1 summarizes the results obtained from the stability analysis and the bifurcation theory approaches.

Table 1. Summary of the properties for von Bertalanffy's functions of polynomial type

	Strong Allee's functions	Weak Allee's functions	Functions with no Allee effect
Possible parameter vals.	$0 < E < K$, $C > -K$ $K > 0, r > 0$	$E = 0$, $C > -K$ $K > 0, r > 0$	$-K < E < 0$, $C > -K$ $K > 0, r > 0$
fixed points	(0) or $(0, A_r, B_r)$	(0) or $(0, A_r, B_r)$	$(0, O_r)$ or $(0, O_r, A_r, B_r)$
Stability $x = 0$	unstable, negative val.	always stable	always unstable
Unconditional extinction reg.	upper bounded by $\Lambda_{(1)_0}$ of fix. pts. A_r and B_r	upper bounded by $\Lambda_{(1)_0}$ of fix. pts. A_r and B_r	not exists this extinction reg.
R_{AE}	$\Lambda_1^* < R_{AE} < \Lambda_{NA}$	$\Lambda_1^* < R_{AE} < \Lambda_{NA}$	not exists R_{AE}
P_{AE}	$P_{AE} \equiv (-K^+, 0^+)$	$P_{AE} \equiv (-K^+, 0^+)$	not exists P_{AE}
P_{BB}	$P_{BB} \equiv (-K^+, 0^+)$	$P_{BB} \equiv (-K^+, 0^+)$	$P_{BB} \equiv (-K^+, 0^+)$

[1] Allee W C 1931 *Animal aggregations* (University of Chicago Press: Chicago)
[2] Avrutin V, Wackenhut G, Schanz M 1999 *Proc. Int. Conf. Tools Math. Modelling* (St. Petersburg) p 4
[3] Avrutin V, Schanz M, Banerjee S 2006 *Nonlinearity* **19** 1875–1906
[4] Avrutin V, Granados A, Schanz M 2011 *Nonlinearity* **24** 2575–98
[5] von Bertalanffy L 1938 *Human Biology* **10** 181–213
[6] Gardini L, Avrutin V, Sushko I 2014 *Int. J. Bifurc. Chaos* **24** 1450024 (30)
[7] Mira C 1987 *Chaotic Dynamics. From the One-Dimensional Endomorphism to the Two-Dimensional Diffeomorphism* (World Scientific: Singapore)
[8] Rocha J L, Aleixo S M 2012 *Discussi. Math., Probab. Statist.* **12** 35–45
[9] Rocha J L, Fournier-Prunaret D, Taha A-K 2014 *Nonlinear Dyn.* **77** (4) 1749–71
[10] Rocha J L, Taha A-K, Fournier-Prunaret D 2015 *Appl. Math. Inf. Sci.* in press

Journal of Physics: Conference Series **692** (2016) 012008 doi:10.1088/1742-6596/692/1/012008

Window of Chaotic Delayed Synchronization

Luís Mário Lopes[1], **Clara Grácio**[2], **Sara Fernandes**[3]

[1]Department of Mathematics, ISEL, Rua Conselheiro Emídio Navarro, 1, 1959-007 Lisboa, Portugal

[2]Department of Mathematics, ECT - Universidade de Évora, CIMA-UE, Rua Romão Ramalho, 59, 7000-585 Évora, Portugal

[3]Department of Mathematics, ECT - Universidade de Évora, CIMA-UE, Rua Romão Ramalho, 59, 7000-585 Évora, Portugal

E-mail: [1]`luismariolopes@gmail.pt`, [2]`mgracio@uevora.pt`, [3]`saf@uevora.pt`

Abstract. We consider a general coupling of two chaotic dynamical systems and we obtain conditions that provide delayed synchronization. We consider four different couplings that satisfy those conditions. We define Window of Delayed Synchronization and we obtain it analytically. We use four different free chaotic dynamics in order to observe numerically the analytically predicted windows for the considered couplings.

Acknowledgment

Research partially sponsored by national funds through the Fundação Nacional para a Ciência e Tecnologia, Portugal-FCT, under the project PEst-OE/MAT/UI0177/2014 and CIMA-UE

1. Introduction

The importance of synchronization is known for a long time and its application is wide [1], [2], [3], [4], [5]. There are different types of synchronization, but the most commonly considered in couplings of discrete dynamical systems is the complete synchronization [6], [7], [8]. Usually, the dynamical systems are supposed to be coupled to others using a linear symmetric coupling, but there are other types of couplings that can be considered. In previous papers, we extended the study of complete synchronization to other types of couplings [9], [10]. Now, we define another type of synchronization, the delayed synchronization. When the delayed synchronization takes place the coupled dynamical systems assume the same values but not at the same time as it happens in the complete synchronization. In the delayed synchronization, one of the coupled system repeats the values of the other after a time delay Δt. Obviously, not all the couplings admit delayed synchronization and even when they admit it they only synchronize with delay for some values of the coupling-strength constant, i.e. only some values of the coupling-strength constant correspond to exponentially stable solutions. Those values define the window of delayed synchronization of the coupling. We are going to define conditions for a coupling to admit delayed synchronization and for the couplings that satisfy those conditions we will be able to obtain the windows of delayed synchronization.

Journal of Physics: Conference Series **692** (2016) 012008 doi:10.1088/1742-6596/692/1/012008

2. Delayed synchronization

We consider couplings of two discrete one-dimensional chaotic dynamical systems of the following type

$$\begin{cases} x\,(t+1) = f\,(x\,(t)) + c \cdot [F_1(x(t)) + F_2(y(t))] \\ y\,(t+1) = g\,(y\,(t)) + c \cdot [G_1(x(t)) + G_2(y(t))] \end{cases} \tag{1}$$

with $c \in [0,1]$. The behavior of the free dynamical systems are chaotic in the sense that $\mu_0 = \lim_{T\to+\infty} \frac{1}{T} \sum_{t=0}^{T-1} \ln |f'|_{s(t)}$ is positive (the same is true of g)

As in the complete synchronization, when there is delayed synchronization the iterates of the dynamical systems of the coupling are identical. Nevertheless, unlike what happens in the complete synchronization the identity occurs for different times t as the following definition stands.

Definition 1. *We say that the coupling (1) admits delayed synchronization of Δt (with $\Delta t \in \mathbb{N}$) if, for each value of the coupling-strength constant c there is a function $s(t)$ such that $(x(t),y(t)) = (s(t+\Delta t),s(t))$ is a solution of (1). If, for each value of c, $s(t)$ can be a solution of a dynamical system with chaotic behavior we say that the complete synchronization has chaotic behavior.*

If the functions f, g, F_1, F_2, G_1 and G_2 satisfy some conditions it is possible to assure that the coupling (1) admits delayed synchronization, such as the following proposition determines.

Proposition 1. *If $f = g$, $F_1 \circ f^{(\Delta t)} + F_2 = 0$ and $G_1 \circ f^{(\Delta t)} + G_2 = 0$, with $f^{(\Delta t)} = f \circ f \circ \ldots \circ f$ (Δt times), then the coupling (1) admits delayed synchronization of Δt with chaotic behavior and the delayed synchronized solution corresponds to $y(t+1) = f(y(t))$*

Proof 1. *If $s(t)$ is a solution of $s(t+1) = f(s(t))$, then $(x(t),y(t)) = (s(t+\Delta t),s(t))$ satisfies (1), for any value of c. In fact, considering, in (1), $f = g$, $F_2 = -F_1 \circ f^{(\Delta t)}$, $G_2 = -G_1 \circ f^{(\Delta t)}$ and $(x(t),y(t)) = (s(t+\Delta t),s(t))$, we obtain*

$$\begin{cases} s(t+\Delta t+1) = f\,(s(t+\Delta t)) + c \cdot \left[F_1(s(t+\Delta t)) - \left(F_1 \circ f^{(\Delta t)}\right)(s\,(t))\right] \\ s\,(t+1) = f\,(s\,(t)) + c \cdot \left[G_1(s(t+\Delta t)) - \left(G_1 \circ f^{(\Delta t)}\right)(s\,(t))\right] \end{cases} \Leftrightarrow$$

$$\begin{cases} s(t+\Delta t+1) = f\,(s(t+\Delta t)) \\ s\,(t+1) = f\,(s\,(t)) \end{cases}$$

Both these equations are obviously verified, since $s(t+1) = f(s(t))$. So, we conclude that the coupling admit a delayed synchronization of Δt. Further, it is a synchronization with chaotic behavior since f corresponds to a chaotic dynamic. □

This proposition determines that a coupling of the following type

$$\begin{cases} x\,(t+1) = f\,(x\,(t)) + c \cdot \left[F_1(x(t)) - F_1(f^{(\Delta t)}(y(t))\right] \\ y\,(t+1) = f\,(y\,(t)) + c \cdot \left[G_1(x(t)) - G_1(f^{(\Delta t)}(y(t))\right] \end{cases} \tag{2}$$

admits a delayed synchronization of Δt with chaotic behavior.

We consider four examples of these type:

• Commanded by the Past Coupling (CPC), corresponding to $\Delta t = 1$, $F_1 = 0$ and $G_1 = id$:

$$\begin{cases} x\,(t+1) = f\,(x\,(t)) \\ y\,(t+1) = f\,(y\,(t)) + c \cdot [x(t) - f(y(t))] \end{cases}$$

(Note that, for the values of c such that the coupling synchronizes, we have $y(t+1) = x(t)$ and x is free, i.e. in that situation y is commanded by the past of the free x; that is why we name this coupling Commanded by the Past Coupling)

• Commanded by the Future Coupling (CFC), corresponding to $\Delta t = 1$, $F_1 = -f$ and $G_1 = 0$:

$$\begin{cases} x(t+1) = f(x(t)) + c \cdot [-f(x(t)) + f(f(y(t)))] \\ y(t+1) = f(y(t)) \end{cases}$$

(Note that, for the values of c such that the coupling synchronizes, we have $x(t) = y(t+1)$ and y is free, i.e. in that situation x is commanded by the future of the free y; that is why we name this coupling Commanded by the Future Coupling)

• Commanded by the Post-Future Coupling (CPFC), corresponding to $\Delta t = 2$, $F_1 = -f$ and $G_1 = 0$:

$$\begin{cases} x(t+1) = f(x(t)) + c \cdot [-f(x(t)) + f(f(f(y(t))))] \\ y(t+1) = f(y(t)) \end{cases}$$

(Note that, for the values of c such that the coupling synchronizes, we have $x(t) = y(t+2)$ and y is free, i.e. in that situation x is commanded by the post-future of the free y; that is why we name this coupling Commanded by the Post-Future Coupling)

• Bidirectional Delayed Coupling (BDC), corresponding to $\Delta t = 1$, $F_1 = -f$ and $G_1 = id$:

$$\begin{cases} x(t+1) = f(x(t)) + c \cdot [-f(x(t)) + f(f(y(t)))] \\ y(t+1) = f(y(t)) + c \cdot [x(t) - f(y(t))] \end{cases}$$

(Note that, unlike the previous couplings, in this one the interaction between the dynamical systems of the coupling is bidirectional and, like that, none of them is free)

3. Window-of-delayed-synchronization

Even if a coupling admits delayed synchronization, only some values of the coupling-strength constant c correspond to an exponentially stable delayed synchronized solution $(x(t), y(t)) = (s(t + \Delta t), s(t))$.

Definition 2. *For a coupling that admits a delayed synchronization of Δt, the window-of-delayed-synchronization (WDS) of the coupling is the set of values of the coupling-strength constant c such that the delayed synchronized solution $(x(t), y(t)) = (s(t + \Delta t), s(t))$ is exponentially stable.*

For the coupling (2) it is possible to analytically obtain the window-of-delayed-synchronization.

Proposition 2. *The window-of-delayed-synchronization of the coupling (2) is*

$$WDS = \{c \in [0,1] : \mu_{ds} < 0\}$$

with

$$\mu_{ds} = \lim_{T \to +\infty} \frac{1}{T} \sum_{t=0}^{T-1} \ln \left| f' \circ f^{(\Delta t)} + c \cdot \left[F_1' \circ f^{(\Delta t)} - G_1' \circ f^{(\Delta t)} \cdot \left(f^{(\Delta t)} \right)' \circ f \right] \right|_{s(t)}$$

and $s(t+1) = f(s(t))$, except eventually by some elements $\{c \in [0,1] : \mu_{ds} = 0\}$.

Journal of Physics: Conference Series **692** (2016) 012008 doi:10.1088/1742-6596/692/1/012008

Proof 2. *In order to simplify the notation, we use x_t instead of $x(t)$ and y_t instead of $y(t)$.*

Considering $u_t = f^{(\Delta t)}(y_t) - x_t$, or equivalently $x_t = f^{(\Delta t)}(y_t) - u_t$, we have around $x_t = f^{(\Delta t)}(y_t)$,

$$\begin{aligned}
u_{t+1} &= f^{(\Delta t)}(y_{t+1}) - x_{t+1} = \\
&= f^{(\Delta t)}\left(f(y_t) + c\cdot\left[G_1(f^{(\Delta t)}(y_t) - u_t) - G_1(f^{(\Delta t)}(y_t)\right]\right) - \\
&\quad -f\left(f^{(\Delta t)}(y_t) - u_t\right) - c\cdot\left[F_1(f^{(\Delta t)}(y_t) - u_t) - F_1(f^{(\Delta t)}(y_t))\right] \simeq \\
&\simeq f^{(\Delta t)}(f(y_t) + c\cdot\left[G_1(f^{(\Delta t)}(y_t)) - G_1'(f^{(\Delta t)}(y_t))\cdot u_t - G_1(f^{(\Delta t)}(y_t))\right]) - \\
&\quad - f\left(f^{(\Delta t)}(y_t)\right) + f'\left(f^{(\Delta t)}(y_t)\right)\cdot u_t - \\
&\quad - c\cdot\left[F_1(f^{(\Delta t)}(y_t)) - F_1'(f^{(\Delta t)}(y_t))\cdot u_t - F_1(f^{(\Delta t)}(y_t))\right] \simeq \\
&\simeq f^{(\Delta t)}\left(f(y_t)\right) - c\cdot\left(f^{(\Delta t)}\right)'\left(f(y_t)\right)\cdot G_1'(f^{(\Delta t)}(y_t))\cdot u_t - \\
&\quad -f\left(f^{(\Delta t)}(y_t)\right) + f'\left(f^{(\Delta t)}(y_t)\right)\cdot u_t + c\cdot F_1'(f^{(\Delta t)}(y_t))\cdot u_t
\end{aligned}$$

So, the linearization of the evolution of u_t is

$$u_{t+1} = \left(f'\left(f^{(\Delta t)}(y_t)\right) + c\cdot\left[F_1'(f^{(\Delta t)}(y_t)) - G_1'(f^{(\Delta t)}(y_t))\cdot\left(f^{(\Delta t)}\right)'\left(f(y_t)\right)\right]\right)\cdot u_t$$

Since for a solution with $x_t = f^{(\Delta t)}(y_t)$ we have $y(t+1) = f(y(t))$, we conclude that if

$$\mu_{ds} = \lim_{T\to+\infty}\frac{1}{T}\sum_{t=0}^{T-1}\ln\left|f'\circ f^{(\Delta t)} + c\cdot\left[F_1'\circ f^{(\Delta t)} - G_1'\circ f^{(\Delta t)}\cdot\left(f^{(\Delta t)}\right)'\circ f\right]\right|_{s(t)} < 0,$$

with $s(t+1) = f(s(t))$, then $u(t) = 0$ is an exponentially stable solution of the previous equation, i.e. $(x(t),y(t)) = (f^{(\Delta t)}(s(t)),s(t)) = (s(t+\Delta t),s(t))$ is an exponentially stable solution of (2) and the corresponding value of c belongs to the window-of-delayed-synchronization. If, instead of that, $\mu_{ds} > 0$, then $u(t) = 0$ is an unstable solution of the previous equation, i.e. $(x(t),y(t)) = (f^{(\Delta t)}(s(t)),s(t)) = (s(t+\Delta t),s(t))$ is an unstable solution of (2) and the corresponding value of c does not belong to the window-of-delayed-synchronization. So, we conclude that $WDS = \{c\in[0,1] : \mu_{ds} < 0\}$, except eventually by some elements $\{c\in[0,1] : \mu_{ds} = 0\}$. □

4. Applications

Now, we obtain the windows-of-delayed-synchronization for the four previously presented couplings applied to four different chaotic systems. Before obtaining them analytically, using the previous proposition, we visualize them numerically using the following procedure:

- we calculate the iterates $(x(t),y(t))$ for t sufficiently large, namely between $t = 100$ and $t = 200$, using random initial conditions [11].
- we use different random initial conditions for each considered value of c.
- we plot $y(t+\Delta t) - x(t)$ as a function of c.

Each graph that is obtained by this procedure show the window-of-complete-synchronization since it corresponds to the set of values of c for which the ordinate is only zero.

We consider four examples of each coupling using four different maps f corresponding to chaotic dynamics . We choose relevant maps with different behavior in what respects to piecewise linearity and number of extrema [12], [13], [14]:

Journal of Physics: Conference Series **692** (2016) 012008 doi:10.1088/1742-6596/692/1/012008

1. The tent map, which is a piecewise linear map with just one maximum:

$$f_1(x) = \begin{cases} 2x, x \in \left[0, \frac{1}{2}\right] \\ 2 - 2x, x \in \left[\frac{1}{2}, 1\right] \end{cases}$$

Its Lyapunov exponent is $\mu_0 = \ln 2$.

2. The logistic map, which is the polynomial interpolation of the tent map using the vertices $x_0 = 0$, $x_1 = \frac{1}{2}$ and $x_2 = 1$ as nodes:

$$f_2(x) = 4x\,(1-x)\,, x \in [0,1]$$

Its Lyapunov exponent is $\mu_0 = \ln 2$.

3. The 3-piecewise linear map, which is a piecewise map with two extrema (one maximum and one minimum):

$$f_3(x) = \begin{cases} 2.4 \cdot x, x \in [0, x_1] \\ 1.7 - 2.4 \cdot x, x \in [x_1, x_2] \\ 2.4 \cdot x - 1.4, x \in [x_2, 1] \end{cases}, \text{ with } x_1 = \frac{17}{48} \text{ and } x_2 = \frac{31}{48}$$

Its Lyapunov exponent is $\mu_0 = \ln 2.4$.

4. The cubic-like map, which is the polynomial interpolation of the 3-piecewise linear map using the vertices $x_0 = 0$, x_1, x_2 and $x_3 = 1$ as nodes:

$$f_4(x) = \frac{x(x-x_2)(x-1)}{x_1(x_1-x_2)(x_1-1)} \cdot 0.85 + \frac{x(x-x_1)(x-1)}{x_2(x_2-x_1)(x_2-1)} \cdot 0.15 + \frac{x(x-x_1)(x-x_2)}{(1-x_1)(1-x_2)}$$

Its Lyapunov exponent is $\mu_0 \simeq 0.715$.

These maps are shown in figure 1.

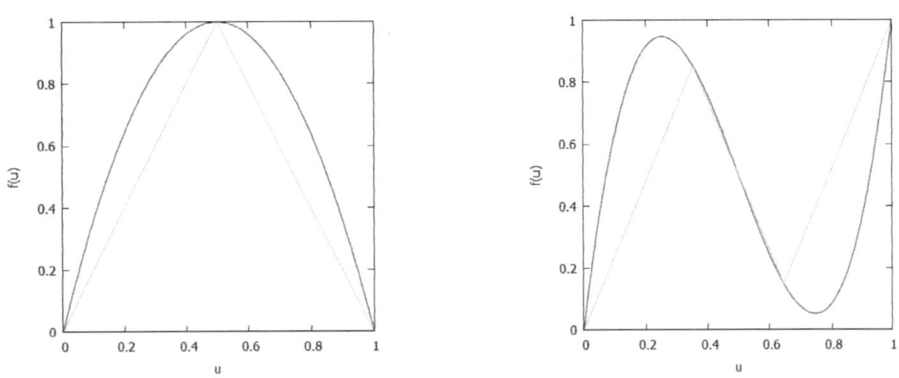

Figure 1: The tent map (left, green), the logistic map (left, blue), the 3-piecewise linear map (right, green) and the cubic-like map (right, blue)

4.1. Windows-of-delayed-synchronization for the Commanded by the Past Coupling

Since the Commanded by the Past Coupling admits a delayed synchronization of 1, we show in figure 2 the $y(t+1) - x(t)$ as a function of c graphs that the numerical procedure provides. We use the four free dynamics considered (corresponding to the tent, logistic,

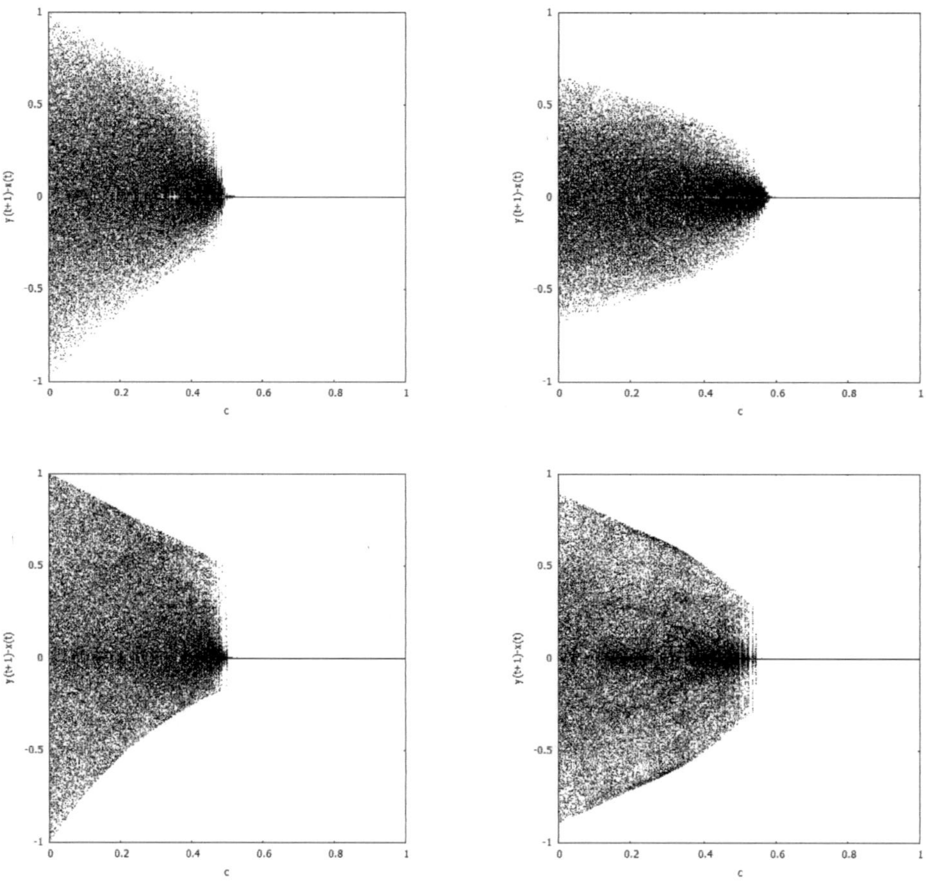

Figure 2: Graphs of the iterates $y(t+1)-x(t)$ as a function of c for the CPC of the tent map (at the top left), the logistic map (at the bottom left), the 3-piecewise linear map (at the top right) and the cubic-like map (at the bottom right)

3-piecewise linear and cubic-like maps) and we verify that all of corresponding couplings have non-empty windows-of-delayed-synchronization.

Along with this, since $\Delta t = 1$, $F_1 = 0$ and $G_1 = id$, we obtain

$$\mu_{ds} < 0 \Leftrightarrow \lim_{T\to+\infty} \frac{1}{T}\sum_{t=0}^{T-1} \ln\left|f' - cf'\right|_{f(s(t))} < 0 \Leftrightarrow \mu_0 + \ln|1-c| < 0 \Leftrightarrow$$

$$\Leftrightarrow 1 - e^{-\mu_0} < c < 1 + e^{-\mu_0} \Rightarrow$$

$$\Rightarrow WDS = \left]1 - e^{-\mu_0}, 1\right]$$

Given the values of the Lyapunov exponents for the chosen maps f, Proposition 2 provides the following windows-of-delayed-synchronization:

· tent and logistic maps: $WDS =]0.5, 1]$

· 3-piecewise linear map: $WDS =]0.58(3), 1]$

· cubic-like map: $WDS \simeq]0.511, 1]$

These results confirm the ones that the numerical approach provides and that the figure 2 shows.

4.2. Windows-of-delayed-synchronization for the Commanded by the Future Coupling

Since the Commanded by the Future Coupling admits a delayed synchronization of 1, we show in figure 3 the $y(t+1)-x(t)$ as a function of c graphs that the numerical procedure provides. We use the four free dynamics considered (corresponding to the tent, logistic, 3-piecewise linear and cubic-like maps) and we verify that all of corresponding couplings have non-empty windows-of-delayed-synchronization.

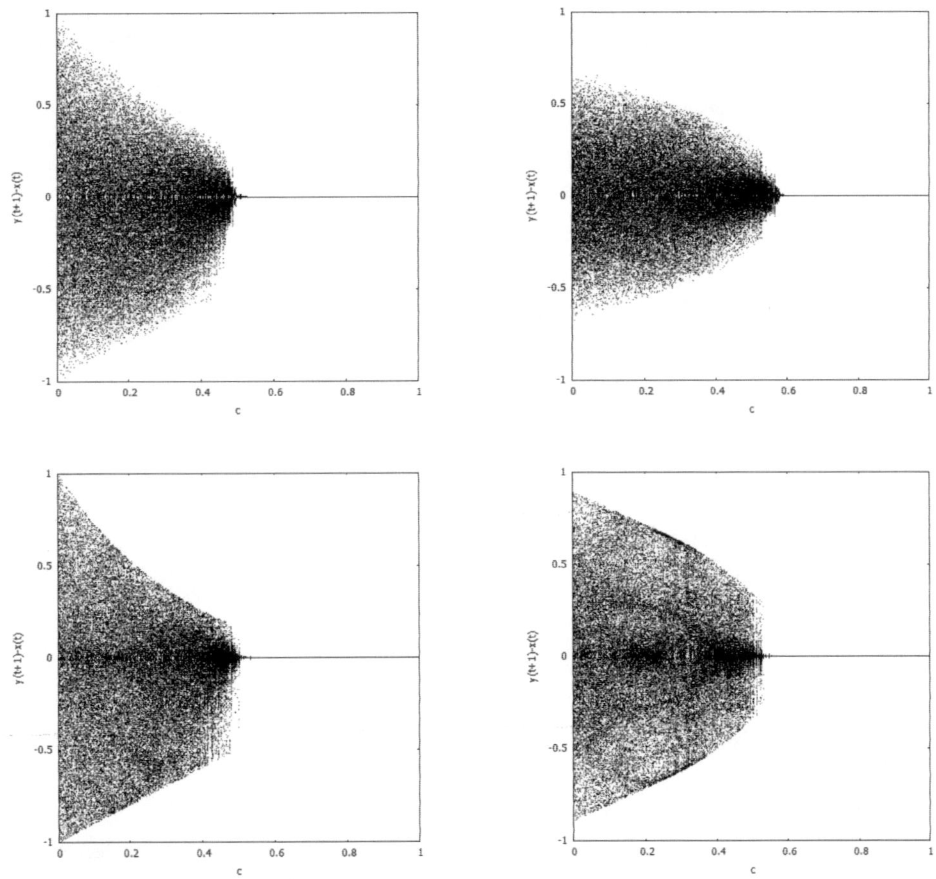

Figure 3: Graphs of the iterates $y(t+1)-x(t)$ as a function of c for the CFC of the tent map (at the top left), the logistic map (at the bottom left), the 3-piecewise linear map (at the top right) and the cubic-like map (at the bottom right)

Along with this, since $\Delta t = 1$, $F_1 = -f$ and $G_1 = 0$, we obtain

$$\mu_{ds} < 0 \Leftrightarrow \lim_{T\to+\infty}\frac{1}{T}\sum_{t=0}^{T-1}\ln\left|f' - cf'\right|_{f(s(t))} < 0 \Leftrightarrow \mu_0 + \ln|1-c| < 0 \Leftrightarrow$$

$$\Leftrightarrow 1 - e^{-\mu_0} < c < 1 + e^{-\mu_0} \Rightarrow$$

$$\Rightarrow WDS = \left]1 - e^{-\mu_0}, 1\right]$$

Given the values of the Lyapunov exponents for the chosen maps f, Proposition 2 provides the following windows-of-delayed-synchronization:

· tent and logistic maps: $WDS =]0.5,1]$
· 3-piecewise linear map: $WDS =]0.58(3),1]$
· cubic-like map: $WDS \simeq]0.511,1]$

These results confirm the ones that the numerical approach provides and that the figure 3 shows.

4.3. Windows-of-delayed-synchronization for the Bidirectional Delayed Coupling

Since the Bidirectional Delayed Coupling admits a delayed synchronization of 1, we show in figure 4 the $y(t+1) - x(t)$ as a function of c graphs that the numerical procedure provides. We use the four free dynamics considered (corresponding to the tent, logistic, 3-piecewise linear and cubic-like maps) and we verify that all of corresponding couplings have non-empty windows-of-delayed-synchronization.

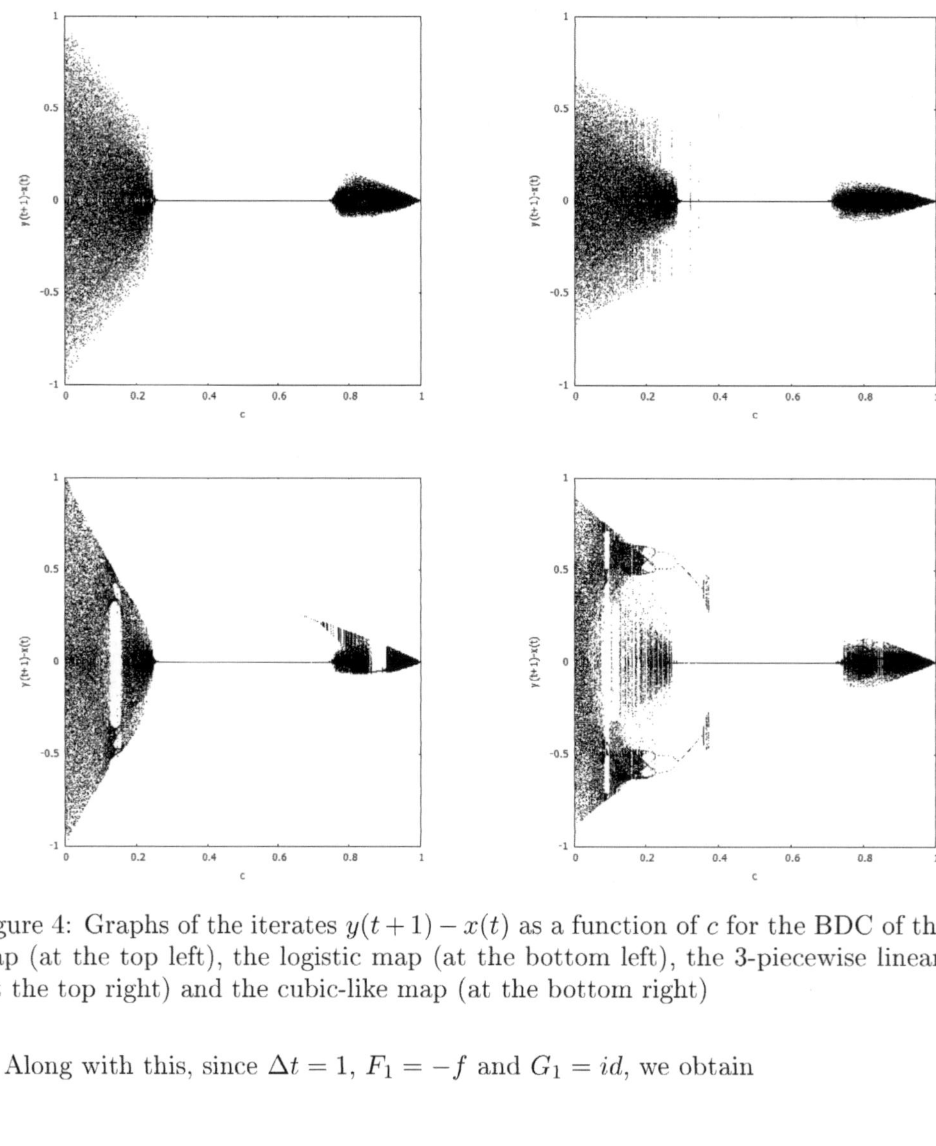

Figure 4: Graphs of the iterates $y(t+1) - x(t)$ as a function of c for the BDC of the tent map (at the top left), the logistic map (at the bottom left), the 3-piecewise linear map (at the top right) and the cubic-like map (at the bottom right)

Along with this, since $\Delta t = 1$, $F_1 = -f$ and $G_1 = id$, we obtain

$$\mu_{ds} < 0 \Leftrightarrow \lim_{T\to+\infty} \frac{1}{T}\sum_{t=0}^{T-1} \ln \left|f' - cf' - cf'\right|_{f(s(t))} < 0 \Leftrightarrow \mu_0 + \ln|1-2c| < 0 \Leftrightarrow$$

$$\Leftrightarrow \frac{1-e^{-\mu_0}}{2} < c < \frac{1+e^{-\mu_0}}{2} \Rightarrow$$

$$\Rightarrow WDS = \left]\frac{1-e^{-\mu_0}}{2}, \frac{1+e^{-\mu_0}}{2}\right[$$

Given the values of the Lyapunov exponents for the chosen maps f, Proposition 2 provides the following windows-of-delayed-synchronization:

· tent and logistic maps: $WDS =]0.25,0.75[$

· 3-piecewise linear map: $WDS =]0.291(6),0.708(3)[$

· cubic-like map: $WDS \simeq]0.2557,0.7443[$

These results confirm the ones that the numerical approach provides and that the figure 4 shows.

4.4. Windows-of-delayed-synchronization for the Commanded by the Post-Future Coupling

Since the Commanded by the Post-Future Coupling admits a delayed synchronization of 2, we show in figure 5 the $y(t+2) - x(t)$ as a function of c graphs that the numerical procedure provides. We use the four free dynamics considered (corresponding to the tent, logistic, 3-piecewise linear and cubic-like maps) and we verify that all of corresponding couplings have non-empty windows-of-delayed-synchronization.

Along with this, since $\Delta t = 2$, $F_1 = -f$ and $G_1 = 0$, we obtain

$$\mu_{sd} < 0 \Leftrightarrow \lim_{T\to+\infty} \frac{1}{T}\sum_{t=0}^{T-1} \ln \left|f' - cf'\right|_{(f\circ f)(s(t))} < 0 \Leftrightarrow \mu_0 + \ln|1-c| < 0 \Leftrightarrow$$

$$\Leftrightarrow 1 - e^{-\mu_0} < c < 1 + e^{-\mu_0} \Rightarrow$$

$$\Rightarrow WDS = \left]1 - e^{-\mu_0}, 1\right]$$

Given the values of the Lyapunov exponents for the chosen maps f, Proposition 2 provides the following windows-of-delayed-synchronization:

· tent and logistic maps: $WDS =]0.5,1]$

· 3-piecewise linear map: $WDS =]0.58(3),1]$

· cubic-like map: $WDS \simeq]0.511,1]$

These results confirm the ones that the numerical approach provides and that the figure 5 shows.

5. Conclusions

We defined delayed synchronization and obtained conditions that assure that a coupling admit it. The delayed synchronized solution is only an exponentially stable one for some values of the coupling-strength constant c, defining an window-of-delayed-synchronization. The windows that we obtained numerically for several couplings that admit delayed synchronization, using several chaotic maps, were confirmed analytically. In fact, we were succeeded in obtaining an analytical expression for the window-of-delayed-synchronization.

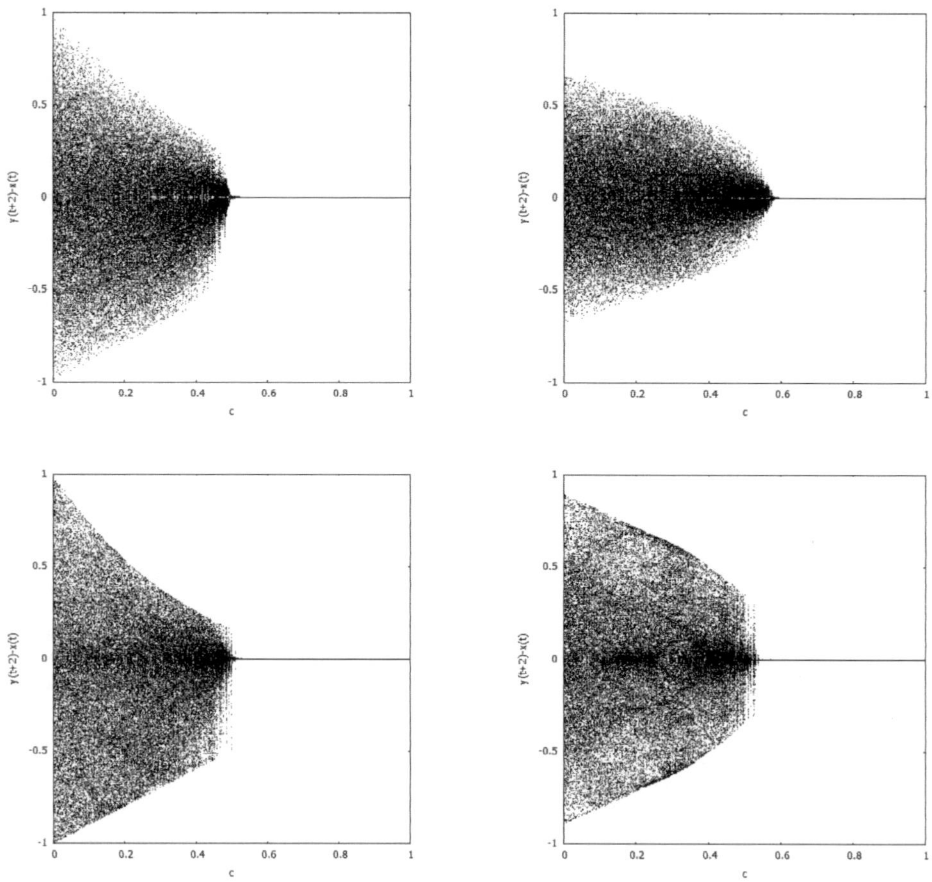

Figure 5: Graphs of the iterates $y(t+2) - x(t)$ as a function of c for the CPFC of the tent map (at the top left), the logistic map (at the bottom left), the 3-piecewise linear map (at the top right) and the cubic-like map (at the bottom right)

References

[1] R. Brown, L. Kocarev, A unifying definition of synchronization for dynamical systems, *Chaos, 10, 344–349*, (2000)

[2] A. Pikovsky, M. Rosenblum, J. Kurths, Synchronization, A Universal Concept in Nonlinear Sciences, *Cambridge University Press, Cambridge,* (2001)

[3] S. Boccaletti, J. Kurths, G. Osipov, D. L. Valladares, C. S. Zhou, The synchronization of chaotic systems, *Phys Rep , 366, 1–101*, (2002)

[4] L.-Q. Chen, A general formalism for synchronization in finite dimensional dynamical systems, *Chaos, Solitons and Fractals, 19, 1239-1242*, (2004)

[5] G. Zhang, Z. R. Liu, Z. J. Ma, Generalized synchronization of different dimensional chaotic dynamical systems, *Chaos, Solitons and Fractals, 32, 2, 773–779*, (2007)

[6] L.M. Pecora, T.L. Carroll, Synchronization in Chaotic Systems, Phys. Rev. Lett., 64, 8, 821-825, (1990)

[7] S. Chen, J. Lu, Parameters identification and synchronization of chaotic systems based upon adaptive control, *Phys. Lett. A, 299, 353-358*, (2002)

[8] G. Zhang, W. Zhang, Z.-R. Liu, Synchronization of Coupled Nonidentical Dynamical Systems, *Chin. Phys. Lett., 27, 3, 030504*, (2010)

[9] L. Lopes, S. Fernandes, C. Grácio, "Windows of Synchronization" and " Non-chaotic Windows", *ESAIM: Proceedings and Surveys, 46, 161-174*, (2014)

[10] L. Lopes, S. Fernandes, C. Grácio, Complete synchronization and delayed synchronization in couplings, *Nonlinear Dynamics, 76, 3, DOI 10.1007/s11071-014-1764-8*, (2014)
[11] A. A. Koronovskii, D. I. Trubetskov, A. E. Khramov, A. E. Khramova,Universal Scaling Laws of Transients, *Doklady Akademii Nauk, 383, 3, 322–325* (2002)
[12] J. Milnor, Remarks on iterated cubic maps, *Experimental Mathematics, 1, 1, 5-24*, (1992)
[13] J. A. Oliveira, E. R. Papesso, E. D. Leonel, Relaxation to Fixed Points in the Logistic and Cubic Maps: Analytical and Numerical Investigation, *Entropy, 15, 4310-4318*, (2013)
[14] J. Gao, B. Sun and J. Yang, The Construction of the Cubic Logistic Chaotic Function Family, *Proc. of Sixth World Congress on Intelligent Control and Automation*, (2006)

Dual Feedback Control in Solid State Lasers: Discrete Maps and Experiments

M.V. Gorbunkov, Yu.Ya. Maslova, Yu.V. Shabalin
P.N. Lebedev Physics Institute, Leninskii prospect 53, Moscow, 119991 Russia

V.G. Tunkin
International Laser Center, Faculty of Physics, Moscow State University, Vorobevy gory, Moscow, 119992 Russia

E-mail: jmaslova@mail.ru

Abstract. Combinations of negative and positive feedback loops were investigated on the basis of discrete map analysis, numerical simulation and experimentally with lamp and diode pumped Nd:YAG lasers. Cases with relative feedback delay in two loops equal to one laser cavity round trip time were considered. Dual feedback control allows to achieve: 1) stable operation of both continuous and pulsed lasers in a wide range of active media gain, 2) laser pulse shortening down to several tens of picoseconds and 3) regular burst mode with periods up to hundred laser cavity round trip times.

1. Introduction

The stabilization of laser operation by single negative feedback loop is well known [1, 2]. Fast inertial negative feedback control allows one to generate short laser pulses via self mode-locking [3, 4]. However sufficient drawback of such a control is a fundamental instability at high active-media gain [5]. Moreover, short pulse duration in self mode-locking regime of Nd:YAG laser is not less than 100 ps [6]. Dual feedback control is a more powerful tool to control the generation dynamics of solid state lasers. In this paper we summarize our experience in dual feedback control of solid state lasers both inside the stability region and above its upper boundary. We focused on the problems of stable operation, laser pulse shortening, and regular bursting with large periods. The advantages of proposed dual feedback control were manifested by discrete map analysis and were verified by numerical simulation and experiments with lamp and diode pumped Nd:YAG lasers. Such control can be widely applied in laser design [7, 8] as well as other areas dealing with optimal control.

2. Highly stable dual feedback controlled laser

For a mode-locked laser controlled by one negative non-inertial (quickly relaxed) feedback loop the recurrent equation is the following:

$$x_{n+1} = rx_n(1 - x_n). \tag{1}$$

Here x_n is the laser pulse energy in the cavity at n-th round-trip, r is the overall gain including active medium gain and passive losses, and the term in brackets represents the feedback loop action which delay is less than a laser cavity round trip time T_r. In steady state $x_{n+1} = x_n = x_{st}$, and we have

$$x_{st} = \frac{r-1}{r}, \qquad \frac{dx_{st}}{dr} = \frac{1}{r^2}. \tag{2}$$

Good operation stability requires the low sensitivity of laser output x to gain variation. From equation (2) we conclude that the sensitivity value dx_{st}/dr decreases when r is increased. Therefore the possibility of steady-state operation at large gain r is the necessary condition for high-level laser stabilization. However, according to [5] maximum acceptable gain for steady operation of a laser controlled by negative feedback with delay $\ll T_r$ equals 3, and for a laser controlled by negative feedback with delay $\sim T_r$ acceptable gain is 2. We have found that dual negative feedback control allows the remarkable broadening of stability region [9]. For a laser controlled by dual non-inertial feedback with relative delay equal to T_r, a recurrence equation is the following:

$$x_{n+1} = rx_n(1 - \alpha x_n - x_{n-1}). \tag{3}$$

Now the term in brackets represents the dual feedback action, α is feedback loops sensitivities ratio and the relative feedback delay is T_r. The regions of stability were found analytically using linear analysis in the neighborhood of x_{st}. The values of multipliers μ [10] were calculated. The stability region covers the region where greater absolute value of multiplier is $\mu < 1$. The upper boundary of stability region r_2 (referred to as the second threshold in literature) corresponds to $\mu = 1$. Figure 1a shows the stability region for map (3) which reveals the point $\alpha_{opt} = 3$ with maximum gain $r_{max} = 5$:

$$r_2(\alpha) = \begin{cases} \alpha + 2, & \alpha \le 3 \\ 3 + \frac{4}{\alpha - 1}, & \alpha > 3 \end{cases} \tag{4}$$

This result is in a contradiction with common opinion [5] that greater delay in a feedback loop always leads to stability decrease.

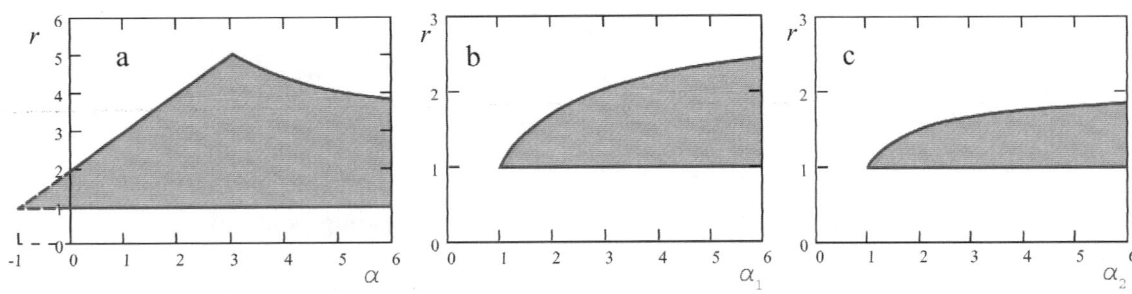

Figure 1. Stability regions for dual feedback controlled systems [9]: a — map (3) controlled by dual non-inertial negative feedback, b — map (5) controlled by negative and delayed positive non-inertial feedback, c — map (6) controlled by positive and delayed negative non-inertial feedback.

It is important to note that combinations of non-inertial positive and negative feedback loops (with negative feedback delay $\ll T_r$ and with positive delay equal to T_r, or with negative feedback delay T_r and with positive delay $\ll T_r$) give no stability increase (figures 1b, c):

$$x_{n+1} = rx_n(1 - \alpha_1 x_n + x_{n-1}), \quad r_2(\alpha_1) = 3 - \frac{4}{\alpha_1 + 1}. \tag{5}$$

$$x_{n+1} = rx_n(1 + x_n - \alpha_2 x_{n-1}), \quad r_2(\alpha_2) = 2 - \frac{1}{\alpha_2}. \tag{6}$$

The recurrence equation of map (3) with two negative feedback loops is close to a system with inertial feedback:

$$x_{n+1} = rx_n \left(1 - \sum_{m=0}^{\infty} x_{n-m}\gamma^m\right). \tag{7}$$

Inertial feedback action is presented by a sum of pulse energies at preceding cavity round-trips taking into account a damping factor γ. Damping factor dependence on feedback memory time τ (later referred to as RC) is given by $\gamma = \exp(-1/\tau)$, where τ is measured in T_r units. Since we discovered stability region broadening for map (3), the stability of systems with inertial control was examined in details. For a laser controlled by two inertial negative feedback loops with relative delay of T_r, the recurrence equation is as follows:

$$x_{n+1} = rx_n \left(1 - \sum_{m=0}^{\infty} x_{n-m}\gamma^m - \alpha_3 \sum_{m=0}^{\infty} x_{n-m-1}\gamma^m\right), \tag{8}$$

where α_3 is the relative sensitivity of feedback loops. To demonstrate stability properties of this two-parametrical (parameters α_3 and γ) system we used multiplier μ diagrams (figures 2–3). Figure 2 shows stability diagrams for map (8). Figure 2a ($\alpha_3 = 0$) corresponds to a laser controlled by a single negative feedback delayed by $\ll T_r$. This system has a narrow stability region (maximum gain is $r_{max} = 2$) and multiplier values close to 1 (which results in long steady-state build-up time).

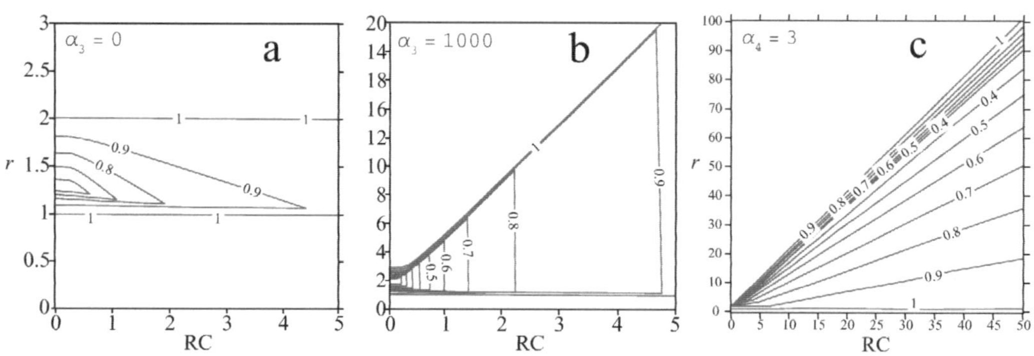

Figure 2. Stability regions with contour plots of greater absolute value of multiplier $\mu = 1, 0.9, \ldots$ [9]: a, b — map (8) controlled by negative feedback (a — case of one inertial negative feedback with delay $\ll T_r$, b — case of one inertial negative feedback delayed by T_r); c — map (9) controlled by negative feedback with delay $\ll T_r$ and positive feedback delayed by T_r.

Figure 2b corresponds to the laser controlled by single-loop negative feedback delayed by T_r. One can see that giant stability region width is achieved at greater RC values. Unfortunately, the absolute value of multipliers increases with RC increase ($\mu \sim 1$). Thus, steady-state build-up time is increased, which results in worse suppression of possible deviations of x from x_{st} that may occur at gain variation.

In contrast to non-inertial systems, combination of positive and negative inertial feedback

$$x_{n+1} = rx_n \left(1 - \alpha_4 \sum_{m=0}^{\infty} x_{n-m}\gamma^m + \sum_{m=0}^{\infty} x_{n-m-1}\gamma^m\right) \tag{9}$$

shows an improvement both in stability region width and multiplier values (figure 2c). Though the maximum acceptable gain value is slightly lower than in the case of two negative feedback

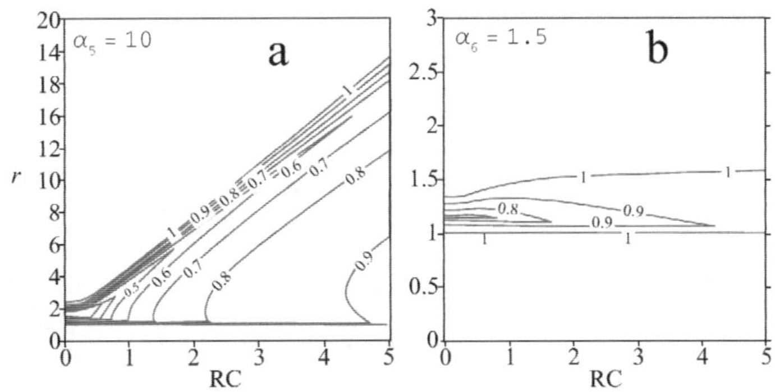

Figure 3. Stability regions with contour plots of greater absolute value of multiplier $\mu = 1, 0.9, \ldots$ [9]: a — system (10) controlled by inertial negative feedback with delay $\ll T_r$ and non-inertial positive feedback delayed by T_r, $\alpha_5 = 10$; b — system (11) controlled by inertial negative feedback delayed by T_r and non-inertial positive feedback with delay $\ll T_r$, $\alpha_6 = 1.5$.

loops, the values of μ are sufficiently smaller, providing deeper stabilization and shorter steady-state build-up times. It is worth to note that low-multiplier region ($\mu < 0.4$) spreads along RC axis for indefinitely great RC values. Our model imposes no restrictions on RC (and r_{max}) increase. Best α_4 values providing optimal stabilization are from 3 to 5. The feedback loops combination described above gives unique capabilities for stabilization. For comparison, the systems controlled by the combination of positive feedback delayed by $\ll T_r$ and negative feedback delayed by T_r, have very narrow stability regions at any α_4. The conclusion is that for the real stabilization less delayed inertial feedback should be always negative. In this case, the second positive feedback delayed by T_r compensates the memory effect of the first negative feedback providing greater stabilization.

Keeping in mind self-mode-locking regime, combination of inertial negative and non-inertial positive feedback loops was also investigated (figure 3):

$$x_{n+1} = rx_n \left(1 - \alpha_5 \sum_{m=0}^{\infty} x_{n-m}\gamma^m + x_{n-1}\right), \tag{10}$$

$$x_{n+1} = rx_n \left(1 - \alpha_6 \sum_{m=0}^{\infty} x_{n-m-1}\gamma^m + x_n\right). \tag{11}$$

Similarly to the case of map (9) when inertial negative feedback was combined with inertial positive feedback, the stability region is broadened and μ remains moderate if a system is controlled by negative feedback with delay $\ll T_r$ and positive non-inertial feedback delayed by T_r (map (10), figure 3a). Combination of negative feedback delayed by T_r and positive feedback with delay $\ll T_r$ gives extremely narrow stability region (map (11), figure 3b).

3. Self-mode-locking in dual feedback controlled laser

Self-mode-locking [3] in a negative feedback-controlled laser by means of electrooptic modulator (EOM) is a well-known technique for stable ultrashort pulses generation. The laser mode-locked by a fast time-shifted negative optoelectronic feedback [11] (figure 4) is a very simple and reliable source of light pulses with several-hundred picosecond duration [4]. In such a laser the losses caused by an intracavity Pockels cell look like a periodic "saw" with a long front and abrupt decay (figure 4c). The decay is formed due to the fast charge of EOM capacity C_m by photocurrent generated in optoelectronic element (OE) under ultrashort light pulse action (figure 4b). The long front is formed by slow discharge of intracavity Pockels cell capacity C_m through a ballast resistor R_0 (figure 4a). This behavior is referred to as a simple relaxator [6].

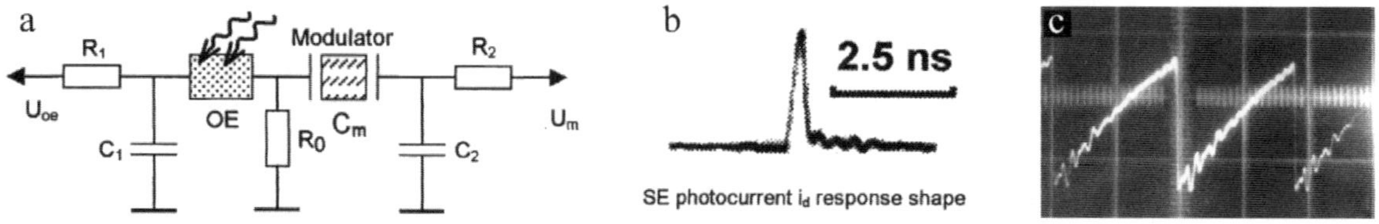

Figure 4. Electrooptic control circuit and signals: a — equivalent scheme, R_0 – ballast resistor, C_m — electrooptic modulator (EOM) capacity, U_m — EOM bias voltage, OE – optoelectronic element, U_{oe} — optoelectronic element bias voltage, C_1, C_2 — capacitors ($C_1 \gg C_2$, $C_2 > C_m$); b — photocurrent pulse; c — EOM control voltage, 5 ns/div.

For self-mode-locking, the discharge time of the EOM capacity should be about T_r [6]. The evolution of the laser field and corresponding control voltage was simulated numerically starting from the spontaneous emission to the ultrashort pulse formation. Finally, the steady-state parameters were calculated. The best conditions for stable self-mode-locking occur when temporal delay in feedback loop corresponds to light pulse passage through the Pockels cell at the moment of the lowest intracavity losses and the maximum EOM transmission. However, this is not the case when the laser is mode-locked by feedback with delay $\ll T_r$ (figure 5a): light pulses are formed on a long front of EOM transmission by slow discharge of the EOM capacity and their duration is inversely proportional to $T_r^{1/2}$ [12, 6, 13]. Therefore, despite fast electronics (i.e. short photocurrent response time) the great value of T_r complicates short pulses generation in lasers controlled by highly stable negative feedback with delay $\ll T_r$.

Some pulse shortening occurs when the pulse passes through the Pockels cell at the moment of the maximum transmission (optimal mode-locking). To achieve this, the delay of a single negative feedback is to be increased up to T_r and the width of the stability region is decreased. We proposed dual feedback combination which is able to overcome the drawbacks of single negative feedback control. It is based on inertial negative feedback with delay $\ll T_r$ and positive quickly relaxed feedback delayed by T_r (figure 6). Positive feedback based on a simple relaxator with fast discharge leads to significant pulse shortening. Figure 7 gives the optical scheme of dual feedback controlled laser system with negative feedback delay $\ll T_r$ and positive feedback delayed by T_r.

The scheme of the picosecond flash-lamp pumped Nd:YAG laser based on the same dual feedback combination is shown in figure 8 [14]. In contrast to the previous scheme, signals reflected from AM and intracavity polarizer were used in feedback loops. Intracavity EOMs were based on $LiTaO_3$ crystals. They were controlled directly by photocurrent of high-voltage subnanosecond silica mesa-structures. Discharge time of the negative feedback loop EOM capacity was as large as $1.5T_r$. Positive feedback response time was not greater than 0.5 ns. For short pulse generation with high stability, optical delay in negative feedback was set as small as possible. Optimal optical delay in positive feedback was about T_r. At the optimal negative and positive feedback sensitivities, the train of short pulses with equal amplitudes and total duration of > 180 μs was generated (figure 9a). Flat top of the train proves the high stabilization properties of the combination of inertial negative feedback loop with delay $\ll T_r$ and fast positive feedback loop delayed by T_r predicted earlier in [9]. Single pulse duration measured by streak-camera was 43.5±1 ps (see figure 9b).

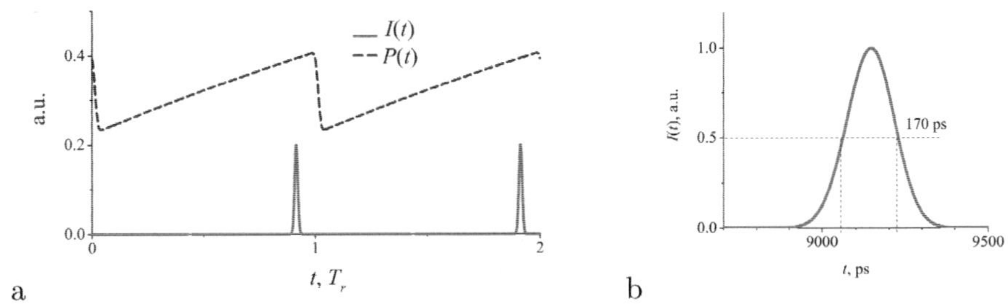

Figure 5. Self-mode-locking regime in a laser controlled by negative feedback with delay $\ll T_r$: $I(t)$ — laser intensity, $P(t)$ — EOM transmission. Cavity round-trip time $T_r = 10$ ns, feedback delay $= T_r/10$, $RC = 2.5T_r$, feedback response time $\tau_i = 0.02T_r$, gain is 100% over the threshold, passive losses coefficient $= 0.1$, EOM bias voltage $U_0 = 0.3U_{\lambda/2}$.

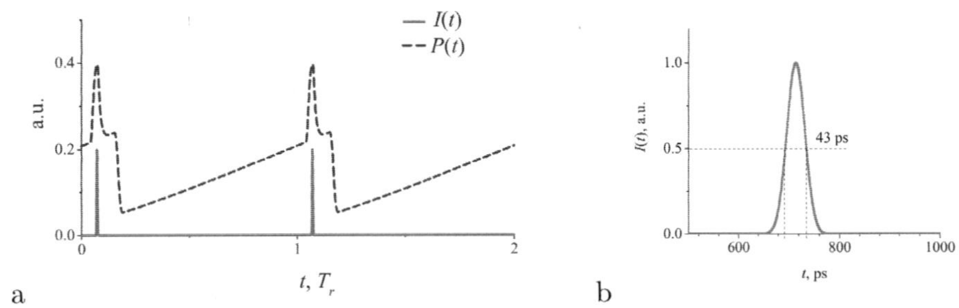

Figure 6. Self-mode-locking regime in a laser controlled by negative feedback with delay $\ll T_r$ and positive feedback delayed by T_r: $I(t)$ — laser intensity, $P(t)$ — EOM transmission. Positive feedback delay is about T_r, $RC = T_r/100$, relative feedback sensitivity is 2.2, feedback response time $\tau_i = 0.02T_r$, other parameters are the same as in figure 5.

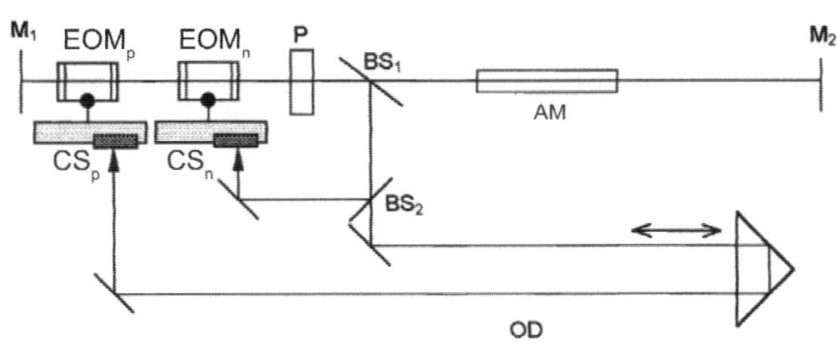

Figure 7. Highly stable dual feedback controlled self-mode-locked laser: M_1, M_2 — cavity mirrors; AM — active medium; BS_1, BS_2 — beam splitters; OD — variable optical delay line for the positive feedback loop; CS_p, CS_n — optoelectronic control circuits of positive and negative feedback loops; EOM_p, EOM_n — electrooptic modulators; P — polarizer. BS_2 provides the optimal positive to negative feedback sensitivity ratio. Optical delay in negative feedback loop (from BS_1 to CS_n) should be as small as possible.

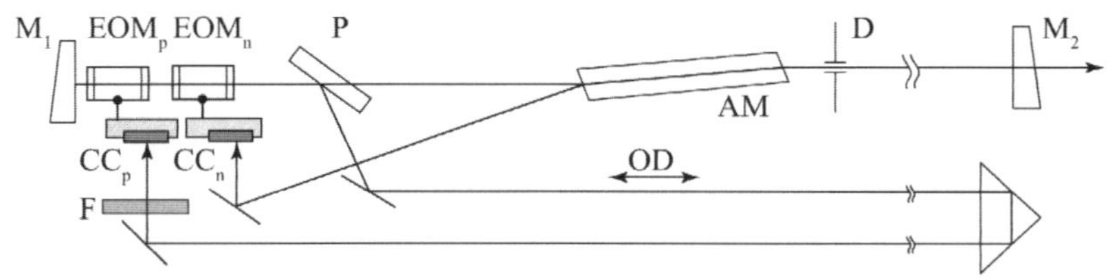

Figure 8. Flash lamp pumped picosecond Nd:YAG laser controlled by positive and negative feedback loops [14]: AM — active medium; M_1, M_2 — cavity mirrors; P — polarizer; D — aperture; CC_n, CC_p — positive and negative feedback control circuits; OD — positive feedback optical delay line; F — neutral density filter.

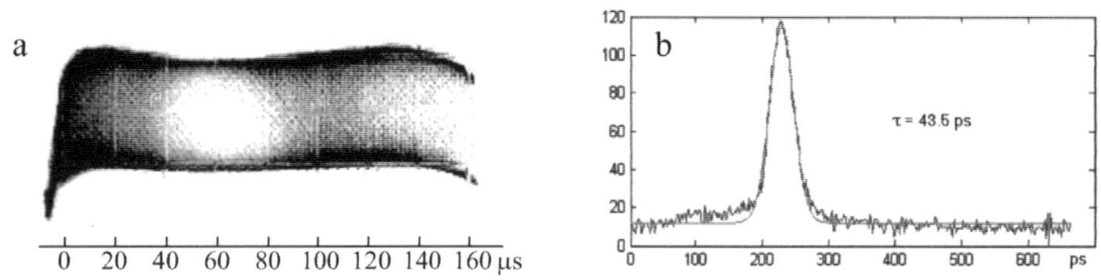

Figure 9. Flash lamp pumped laser output under dual feedback control: a — train of picosecond pulses; b — short laser pulse measured by streak-camera and its fit (duration 43.5 ps) [14].

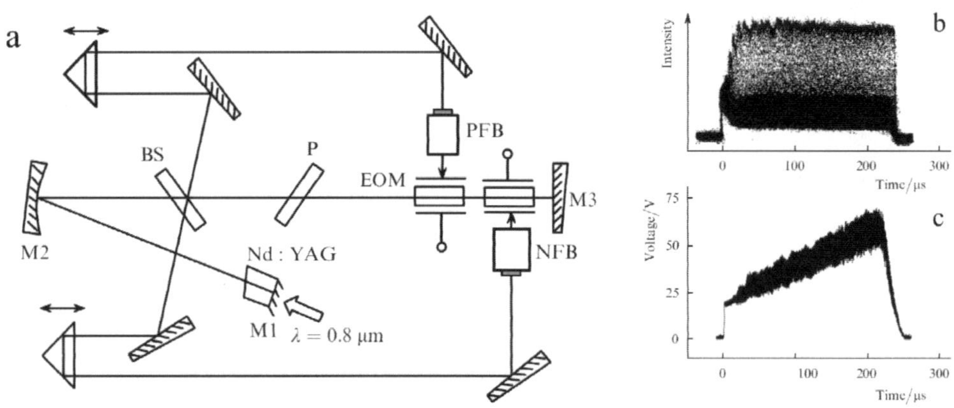

Figure 10. Diode end-pumped picosecond Nd:YAG laser controlled by dual feedback [7]: a — optical scheme: M1–M3 — mirrors, BS — beamsplitter, P — polariser; b — laser output radiation; c — negative feedback voltage.

Later this approach was implemented in a diode pumped Nd:YAG laser (figure 10) [7]. The laser cavity of length 1.2 m consisted of highly reflecting mirror M1 deposited on the flat end face of the AM, two highly reflecting mirrors (spherical mirror M2 with a radius of curvature 2 m and flat mirror M3). AM was 15 mm long and 5 mm in diameter, a low-voltage bisectional EOM was based on $LiTaO_3$ crystals. Radiation to control laser operation was deflected from the cavity using beam splitter BS. After optical delay line, the radiation was directed onto the fast semiconductor structures in the negative and positive feedback circuits. In spite of the considerable variation of AM gain in the course of pulsed diode pumping, effective stabilization of radiation led to the formation of picosecond pulse trains with nearly constant amplitude of duration exceeding 200 μs (figure 10b). The time dependence of the negative feedback voltage is shown in figure 10c. The duration of picosecond pulses measured by AGAT SF-3 streak camera was 30 ps. The laser pulse generated by dual feedback-controlled laser system proved to be more than 3 times shorter than in single-loop feedback-controlled laser. This improvement is paid by the cost of self-mode-locking region narrowing.

4. Regular nonlinear dynamics with large period

In spite of significant shrinking of the stability region width in the case when positive feedback delay is shorter than the negative feedback delay (figures 1c, 3b), this combination is useful in generation of regular bursts with large periods compared to T_r. The period of regular dynamics in map (3) calculated at the stability region boundary depends on the relative sensitivity of feedback loops α, and in the region of positive and negative feedback control (dashed line in figure 1, $0 \leq \alpha < 1$) is approximated by [16]:

$$T(r_2(\alpha)) = \frac{2\pi}{\sqrt{\alpha+1}}. \tag{12}$$

To realize the idea, a laser system was assembled in which a signal reflected from an intracavity Pockels cell polarizer is directed to an optoelectronic negative feedback circuit. In such laser the control scenario corresponds to the combination of positive and delayed negative feedback loops (see figure 11) [16]:

$$y_{n+1} = y_n r \left(1 - y_{n-1} + \frac{P_0}{r} y_n\right), \tag{13}$$

where P_0 is the initial electrooptic modulator transmission. We take into account two-pass Pockels cell transmission

$$P_0(U_m) = \cos^2\left(\frac{U_m}{U_{\lambda/4}}\frac{\pi}{2}\right), \tag{14}$$

where U_m is a static bias voltage (figure 4) and $U_{\lambda/4}$ is a cell quarter-wave voltage. Applying (12) we estimate the nonlinear dynamics development threshold gain $r_2(U_m)$ and oscillation period $T(U_m)$ in two limits. If $U_m \longrightarrow U_{\lambda/4}$ then $T(U_m) \longrightarrow 6$ and $r_2(U_m) \longrightarrow 2$. When U_n is far less than $U_{\lambda/4}$,

$$T(U_m) \approx 2\sqrt{\frac{2\pi U_{\lambda/4}}{U_m}}. \tag{15}$$

Therefore, in a laser controlled by optoelectronic negative feedback with the control signal obtained from the Pockels cell polarizer, regular bursts with periods starting from 6 up to hundreds T_r can be obtained above the stability region boundary. Such regime can be applied in master oscillator of optical unit in a pulsed laser-electron Thomson X-ray source [8]. This idea was implemented using a flash-lamp pumped Nd:YAG laser ($\oslash 6.3 \times 60$ mm rod was used) [16]. A PC-controlled laser pumping unit based on the incomplete discharge of large capacity allowed us to vary pump duration up to 3.9 milliseconds and pump energy up to 250 J. Laser cavity consisted of flat wedge-shaped mirrors of 98 and 35% reflection ($\lambda = 1064$ nm). An intracavity mirror telescope 3:1 allowed to enlarge a laser mode volume in AM and thus to raise the laser output. Total cavity length was 150 cm. An iris aperture was used for mode selection. A two-sectional temperature-compensated Pockels cell based on two $3 \times 3 \times 15$ mm^3 $LiTaO_3$ crystals ($U_{\lambda/4} \sim 200$ V for $\lambda = 1064$ nm) and a multilayer polarizer (Brewster angle) was placed close to the rear laser mirror. Variable bias voltage U_m was applied to the Pockels cell. A high voltage (1200–1300 V) sub-nanosecond photocurrent response silicon mesa-structure [6] was used as a

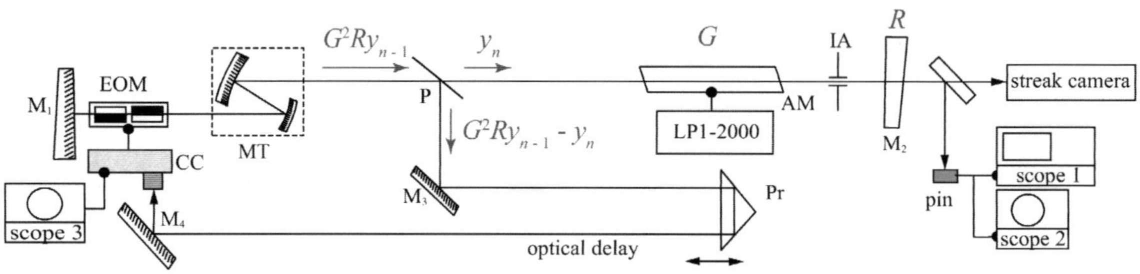

Figure 11. Millisecond lamp pumped Nd:YAG laser designed for submicrosecond scale bursts [16]: AM — active laser medium; P polarizer; M_1, M_2 — cavity mirrors; M_3, M_4 — optical delay mirrors; Pr — prism; IA — iris aperture; MT — mirror telescope; EOM — electrooptic modulator based on $LiTaO_3$ crystals; CC — feedback control circuit; G — AM gain, R — mirror reflectivity, y_n — normalized laser pulse energy at the n-th round-trip.

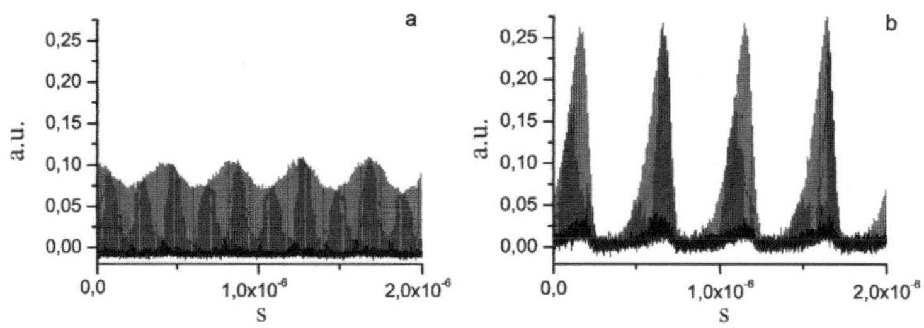

Figure 12. Oscilloscope traces of laser output, signal is proportional to picosecond pulse energy [16]. The development of regular 0.5 microsecond bursts: a — harmonic modulation, b — developed bursts.

control element of the optoelectronic system. Discharge time of intracavity Pockels cell capacity was set to 20 ns ($2T_r$). The control signal was taken from an intracavity Pockels cell polarizer (figure 11). Setting the optimal feedback time delay and the diameter of intracavity iris aperture we obtained a train of 100-picosecond pulses with slowly decreasing amplitude and total number of pulses up to 350000. By increasing pump power we observed regimes of regular bursts with controlled period from 25 up to $75T_r$ as U_m was decreased from $0.3U_{\lambda/4}$ to $0.05U_{\lambda/4}$. An example of 0.5 μs period bursting development is shown in figure 12a, b (microsecond time scale). Period of the developed bursts (figure 12b) is $\leq 25\%$ greater than the period of harmonic modulation under the same U_m. The idea to use the control signal reflected from the polarizer was realized in diode-pumped laser as well [17].

5. Conclusions

The investigation of laser operation inside the stability region showed that dual feedback control provides stabilization improvement. To achieve this, inertial negative feedback with delay $\ll T_r$ is combined with positive feedback delayed by T_r. In such cases stabilization improvement is determined by: 1) non-sensitivity to gain variation owing to the broadening of the stability region and 2) realization of suitable multiplier absolute values. Low values of multipliers provide deep stabilization and shorter steady-state build-up time. Optimal positive to negative feedback sensitivities ratio is 10–50%. In self-mode-locked laser, positive feedback provides not only greater stability but also about 3 times pulse shortening due to the abrupt loss modulation.

Above the upper boundary of the stability region, the period of regular bursts can be increased by means of feedback combination when the positive feedback acts at the time T_r earlier than the negative one. Such combined action was implemented in the laser system controlled by optoelectronic feedback which uses the signal from the intracavity Pockels cell polarizer. The period is increased when the Pockels cell bias voltage is decreased. Burst periods up to $90T_r$ were obtained.

Acknowledgments
We are grateful to D.V. Yakovlev, V.B. Morozov, A.N. Olenin, L.S. Telegin, V.A. Petukhov, M.A. Semenov, A.V. Vinogradov, P.V. Kostryukov, and K.A. Bubnov for fruitful discussions. This work was supported by the Fundamental research program of Presidium RAS "Fundamental and applied photonics problems and new optical material physics", and by the Scientific Educational Complex Program of P.N. Lebedev Physics Institute RAS.

References
[1] F.R. Marshall, D.L. Roberts,"Use of Electro-Optical Shutters to Stabilize Ruby Laser Operation", Proceedings of the IRE, 50, 2108, 1962.
[2] A.S. Agabekyan, A.Z. Grasyuk et al., "Stabilization of unstable regime in a two-level quantum generator", Sov. Radiotechnics and Electronics, 9, No. 12, 2156–2165, 1964.
[3] V.K. Makukha, V.S. Smirnov, V.M. Semibalamut, "Generation of ultrashort pulses in a negative-feedback laser", Sov. J. Quantum Electron., 7, pp. 573–575, 1977 (doi:10.1070/QE1977v007n05ABEH012561).
[4] D.B. Vorchik, M.V. Gorbunkov, "Self mode locking in Nd glass laser by fast negative feedback with solid state optoelectronic system", Bulletin of the Lebedev Physics Instituite, 12, pp. 70–76, 1997.
[5] I.M. Bayanov, V.M. Gordienko, M.G. Zvereva, S.A. Magnitskii, A.P. Tarasevich, "Highly stable picosecond YAG:Nd^{3+} laser with a negative feedback", Sov. J. Quantum Electron., 19, pp. 994–996, 1989 (doi:10.1070/QE1989v019n08ABEH008662).
[6] M.V. Gorbunkov, Yu.V. Shabalin, "Picosecond YAG:Nd3 Laser with Self-Excitation of HF Oscillations in Optoelectronic Negative Feedback System", Bulletin of the LPI, 8, 38–50, 1998.
[7] M.V. Gorbunkov, A.V. Konyashkin, P.V. Kostryukov et al., "Pulsed-diode-pumped, all-solid-state, electro-optically controlled picosecond Nd:YAG lasers", Quantum Electronics, 35(1), 2-6, 2005 (doi: 10.1070/QE2005v035n01ABEH002893).
[8] M.V. Gorbunkov, Yu.Ya. Maslova, A.V. Vinogradov, "Optical unit of Laser-Electron X-ray Generator designed for medical applications", Nuclear Instruments and Methods A 608, S32–S35, 2009.
[9] M.V. Gorbunkov, Yu.V. Shabalin, "Two-Loop Feedback Controlled Laser: New Possibilities for Ultrashort Pulses Generation and High-Level Stabilization", Proc. SPIE, 4751, 463, 2002.
[10] Yu.A. Kuznetsov, *Elements of Applied Bifurcation Theory*, Second Edition, 1998.
[11] D.B. Vorchik, M.V. Gorbunkov, "Optoelectronic control of solid state lasers using new high voltage silicon elements", in Proc. CLEO/Europe'96, Hamburg, Germany, 282 1996.
[12] D. Kuizenga, A. Siegman, "FM and AM mode locking of the homogeneous laser - Part I: Theory", IEEE Journal of Quantum Electronics, 6, 694-708, 1970.
[13] M.V. Gorbunkov, Yu.V. Shabalin, "Laser Pulse Shortening Due to Self-Excited HF Oscillations in Optoelectromc Negative Feedback System", Proceedings of SPIE, 3516, 288–293, 1998.
[14] M.V. Gorbunkov, V.B. Morozov, A.N. Olenin, Yu.V. Shabalin, L.S. Telegin, V.G. Tunkin, D.V. Yakovlev, "Modern approaches to the pulsed diode-pumped picosecond lasers design", 5th Italian-Russian Laser Symposium ITARUS 2003, October 29-November 2, 2003, Moscow, Russia, Technical Digest, 34–37.
[15] M.V. Gorbunkov, Yu.Ya. Maslova, Yu.V. Shabalin, A.V. Vinogradov, in WDS'07 Proceedings of Contributed Papers: Part III - Physics (eds. J. Safrankova and J. Pavlu), Prague, Matfyzpress, pp. 140–144, 2007.
[16] M.V. Gorbunkov, Yu.Ya. Maslova, V.A. Petukhov, M.A. Semenov, Yu.V. Shabalin, A.V. Vinogradov, "Submicrosecond regular and chaotic nonlinear dynamics in a pulsed picosecond Nd:YAG laser with millisecond pumping", Applied Optics, Vol. 48, Issue 12, pp. 2267–2274, 2009.
[17] E.G. Bessonov, K.A. Bubnov, A.V. Vinogradov et al., "Prototype of optical system of laser-electron X-ray source for medical applications". Proceedings of 5th Russian school for students and young scientists in laser physics and laser technology, Sarov, 26–29 April 2011.

Journal of Physics: Conference Series **692** (2016) 012010 doi:10.1088/1742-6596/692/1/012010

Stability as a Whole of a Family of Fibers Maps and Ω-Stability of C^1-Smooth Skew Products of Maps of an Interval

Lyudmila S. Efremova

Institute of Information Technologies, Mathematics and Mechanics, University of Nizhni Novgorod, Gagarin ave., 23a, Nizhni Novgorod, 603950, Russia

E-mail: `lefunn@gmail.com`

Abstract. Properties of C^1-smooth skew products of maps of an interval with stable as a whole family of fibers maps are established. These results are applied to the proof of the criterion of Ω-stability of C^1-smooth skew products of maps of an interval (with respect to homeomorphisms of skew products class).
The proper subspace of the space of C^1-smooth skew products of maps of an interval is distinguished, where Ω-stable C^1-smooth skew products are contained. It is proved that Ω-stable skew products are not dense in the distinguished subspace of C^1-smooth maps.

1. Introduction

Different aspects of C^1- structural stability and C^1- Ω-stability are investigated in works [1 – 9] for diffeomorphisms and flows and in works [10 – 12] for endomorphisms.
Some properties of fibers maps of C^1-smooth Ω-stable skew products of maps of an interval (with respect to homeomorphisms of skew products class) are investigated in papers [13 – 16]. The concept of stability as a whole of a family of fibers maps for a C^1-smooth skew product of maps of an interval is introduced in [15].
This work is the direct continuation of papers [13 – 18]. We solve here the nonlocal problem of description of the set of C^1-smooth Ω-stable skew products of maps of an interval. First, with the use of the concept of stability as a whole in C^1-norm of a family of fibers maps we prove the criterion of C^1- Ω-stability (with respect to homeomorphisms of skew products class). Second, using the decomposition theorem for the space of C^1-smooth skew products of maps of an interval with a complicated dynamics of a quotient map (see [17, 18]), we distinguish the proper subspace of this space, where C^1-smooth Ω-stable skew products of maps of an interval with a complicated dynamics of a quotient map are contained. Third, we prove that Ω-stable C^1-smooth skew products are not dense in the mentioned above proper subspace. This last result strengthens the main result of paper [16], where nondensity of Ω-stable C^1-smooth skew products of maps of an interval is proved in the space of all C^1-smooth skew products of maps of an interval.

Let $I = I_1 \times I_2$ be a closed rectangle in the plane (I_1, I_2 are closed intervals). We consider a skew product of maps of an interval, i. e. a dynamical system $F : I \to I$, where

$$F(x, y) = (f(x), g_x(y)), \quad \text{and} \quad g_x(y) = g(x, y), \ (x; y) \in I. \tag{1}$$

Journal of Physics: Conference Series **692** (2016) 012010 doi:10.1088/1742-6596/692/1/012010

The map $f : I_1 \to I_1$ is called *the quotient map* of skew product (1), and the map $g_x : I_2 \to I_2$ is called the map *acting in the fiber over an arbitrary point* $x \in I_1$.
By formula (1) the equality

$$F^n(x, y) = (f^n(x), g_{x,n}(y)), \quad \text{where} \quad g_{x,n} = g_{f^{n-1}(x)} \circ \ldots \circ g_x, \tag{2}$$

is valid for every natural number n and every point $(x; y) \in I$.

Let, as usually, $T^0(I)$ ($T^1(I)$) be the space of all continuous (all C^1-smooth) skew products of maps of an interval with the standard C^0-norm (the standard C^1-norm).
Denote by $C^1_{\partial_k}(I_k)$ ($k = 1, 2$) the subspace of the space $C^1(I_k)$ (of C^1-smooth maps of the segment I_k into itself with the standard C^1-norm), which consists of all maps $\psi \in C^1(I_k)$ satisfying the condition of ψ-invariance of the boundary ∂I_k of the segment I_k:

$$\psi(\partial I_k) \subseteq \partial I_k.$$

Let us remind that a map $\xi \in C^1_{\partial_k}(I_k)$ ($k = 1, 2$) is Ω-stable in C^1-norm (i. e. in the space $C^1_{\partial_k}(I_k)$) if for every $\delta > 0$ there exists $\varepsilon > 0$ such that for every map $\varphi \in B^1_{k,\varepsilon}(\xi)$ one can find δ-closed in C^0-norm to the identity map homeomorphism $h : \Omega(\xi) \to \Omega(\varphi)$ satisfying the equality

$$h \circ \xi|_{\Omega(\xi)} = \varphi|_{\Omega(\varphi)} \circ h, \tag{3}$$

where $B^1_{k,\varepsilon}(\cdot)$ is ε-neighborhood of a map in the space $C^1_{\partial_k}(I_k)$ (with respect to C^1-norm); $\Omega(\cdot)$ is the nonwandering set of a map[1].
Denote by $C^1_\omega(I_k)$ the space of all Ω-stable in $C^1_{\partial_k}(I_k)$ maps of the closed interval I_k into itself ($k = 1, 2$).

Proposition 1 [10, 20]. *Let $f \in C^1_\omega(I_1)$. Then*

(1.1) *either f is a map of type $\prec 2^\infty$ (i. e. the set of the (least) periods of f-periodic points coincides with the set $\{1, 2, \ldots, 2^\mu\}$ for some $0 \leq \mu < +\infty$)), and in this case the nonwandering set $\Omega(f)$ is finite and consists of hyperbolic periodic points;*

(1.2) *or f is a map of type $\succ 2^\infty$ (i. e. there exists an f-periodic point x ($x \in Per(f)$) with the (least) period $n(x) \notin \{2^i\}_{i \geq 0}$), and in this case the nonwandering set $\Omega(f)$ is the union of finitely many hyperbolic periodic points and finitely many locally maximal quasiminimal sets[2], which are hyperbolic, perfect and nowhere dense. ("Locally maximal" means "maximal in a neighborhood of itself".)*

The set $C^1_\omega(I_1)$ is open and everywhere dense in $C^1_{\partial_1}(I_1)$.

Define the space $T^1_*(I)$ of C^1-smooth skew products of maps of an interval (with the standard C^1-norm) as the subspace of the space $T^1(I)$, which consists of skew products of maps of an interval with quotients from the space $C^1_\omega(I_1)$. As it follows from Proposition 1, the set $T^1_*(I)$ is open and everywhere dense in the subspace of the space $T^1(I)$ consisting of skew products with quotient maps from $C^1_{\partial_1}(I_1)$.

Results of this paper are obtained with the use of the special multifunctions related to an arbitrary continuous skew product of maps of an interval.

Definition 1 [13]. *The Ω-function of a map $F \in T^0(I)$* is the multifunction $\zeta^F : \Omega(f) \to 2^{I_2}$ satisfying the equality

$$\zeta^F(x) = (\Omega(F))(x)$$

for any $x \in \Omega(f)$, where $(\Omega(F))(x) = \{y \in I_2 : (x; y) \in \Omega(F)\}$ is the slice of the nonwandering set $\Omega(F)$ by the vertical fiber over a point x, 2^{I_2} is the topological space of closed subsets of I_2 with the exponential topology [22, 23].

[1] Definitions of wandering, nonwandewring points and sets one can find in [19].
[2] A quasiminimal set of a map is the closure of an infinite recurrent trajectory [21].

Journal of Physics: Conference Series **692** (2016) 012010 doi:10.1088/1742-6596/692/1/012010

The following definition will be given here for a skew product $F \in T^1_*(I)$. Let n be a natural number. By Proposition 1 the equality $\Omega(f^n) = \Omega(f)$ holds in considering case.
Let us use the skew product

$$F_n(x,\, y) = (id(x),\, g_{x,\,n}(y))$$

and the direct product

$$F_{n,1}(x,\, y) = (f^n(x),\, id(y)),$$

where $id(x)$ and $id(y)$ are the identity maps of the closed intervals I_1 and I_2, respectively, and $F_n,\, F_{n,\,1} : I \to I$. Then the following formula is valid:

$$F^n = F_{n,\,1} \circ F_n. \tag{4}$$

Formula (4) allows to define new multifunctions for each iteration of F. The graphs of these multifunctions are using to form the nonwandering set of the map $F \in T^1_*(I)$ (or, equivalently, to form the graph of the Ω-function of F).

Definition 2. *An auxiliary multifunction of a map* $F \in T^1_*(I)$ is a multifunction $\eta_n : \Omega(f) \to 2^{I_2}$ satisfying

$$\eta_n(x) = \Omega(g_{x,\,n})$$

for any $x \in \Omega(f)$ (see [13]), where $\Omega(g_{x,\,n})$ is the nonwandering set of a map $g_{x,\,n} : I_2 \to I_2$.
A function $\overline{\eta}_n : \Omega(f) \to 2^{I_2}$ is said to be *a multifunction suitable to the Ω-function of a map* $F \in T^1_*(I)$ if the graph of $\overline{\eta}_n$ in I is the closure of the graph of the auxiliary function η_n. We have

$$\overline{\eta}_n(x) = (\overline{\eta}_n)(x) \quad \text{for any } x \in \Omega(f)$$

(see [17, 18]). Here $(\overline{\eta}_n)(x)$ denotes the slice of the graph of $\overline{\eta}_n$ by the fibre over x (or, equivalently, the slice of the closure of the graph of η_n).

Having defined auxiliary functions η_n (suitable functions $\overline{\eta}_n$) for all $n > 1$, we must move each point $(x;\, y)$ on the graph of η_n (on the graph of $\overline{\eta}_n$, respectively) to the point $(f^n(x);\, y)$ using the direct product $F_{n,\,1}$ (see equality (4)). We can therefore define multifunctions $\eta_{n,\,1} : \Omega(f) \to 2^{I_2}$ ($\overline{\eta}_{n,\,1} : \Omega(f) \to 2^{I_2}$), $n > 1$, by the equalities

$$\eta_{n,\,1}(x) = (F_{n,\,1}(\eta_n))(x) \quad (\overline{\eta}_{n,\,1}(x) = (F_{n,\,1}(\overline{\eta}_n))(x)$$

for any $x \in \Omega(f)$. Here η_n ($\overline{\eta}_n$) is the graph of the corresponding multifunction in I, and $(F_{n,\,1}(\eta_n))(x)$ ($(F_{n,\,1}(\overline{\eta}_n))(x)$) is the slice of the set $F_{n,\,1}(\eta_n)$ (of the set $F_{n,\,1}(\overline{\eta}_n)$) by the fibre over $x \in \Omega(f)$. Since any point $(x;\, y)$ on the graph of $\eta_{n,\,1}$ (on the graph of $\overline{\eta}_{n,\,1}$) can be reached from any point $(\overline{x};\, y)$ on the graph of η_n (on the graph of $\overline{\eta}_n$) using $F_{n,\,1}$, where $\overline{x}$ is a point in the n-th complete preimage of x under the map $f|_{\Omega(f)}$, it follows that

$$\eta_{n,1}(x) = \bigcup_{\overline{x} \in \{f^{-n}(x)\}} \eta_n(\overline{x}) \qquad \Big(\overline{\eta}_{n,1}(x) = \bigcup_{\overline{x} \in \{f^{-n}(x)\}} \overline{\eta}_n(\overline{x})\Big)$$

We consider the subspace of the space $T^1_*(I)$, which consists of skew products with quotient maps of type $\succ 2^\infty$. Following [17, 18] we distinguish four main subspaces of this space.
Let $F \in T^1_*(I)$ be a skew product with a quotient map of type $\succ 2^\infty$. Then by Proposition 1 the perfect part of the nonwandering set of its quotient map $\Omega_p(f)$ is not empty. Let $K(f) \subset \Omega_p(f)$ be a locally maximal quasiminimal set of f, and $\tau(f_{|K(f)})$ be the set of the (least) periods of

periodic points of $f_{|K(f)}$. There exist natural numbers $m_0 = m_0(K(f))$, $i_0 = i_0(K(f))$ and a finite subset $N_* = N_*(K(f))$ of the set $\mathbf{N}$ of natural numbers (possibly, empty) such that[3]

$$\tau(f_{|K(f)}) = \{m_0 i\}_{i \geq i_0} \cup N_* \text{ (see [25, 26]).}$$

We need the following natural numbers:

$$m_* = \underset{K(f) \subset \Omega_p(f)}{\text{l.c.m.}} \{m_0(K(f))\},$$
$$n_* = \underset{K(f) \subset \Omega_p(f)}{\text{l.c.m.}} \{n \in N_*(K(f))\},$$
$$i_* = \max_{K(f) \subset \Omega_p(f)} \{i_0(K(f))\},$$

where l.c.m. is the least common multiple of a finite set of natural numbers[4].

Definition 3 [17, 18]. We say that a skew product $F \in T^1_*(I)$ with a quotient map of type $\succ 2^\infty$ *satisfies condition* **H** *(strong condition* **H***)* if for any sequence of natural numbers $\{l^*_i\}_{i \geq i_*}$ with

$$l^*_i = m_* n_* i, \tag{5}$$

the multifunctions $\overline{\eta}_{l^*_i}$ (the multifunctions $\eta_{l^*_i}$ respectively) are continuous for all $i \geq i^*$, where $i^* \geq i_*$.

As it follows from the definition of multifunctions $\overline{\eta}_{l^*_i,1}$ (and $\eta_{l^*_i,1}$) continuity of a multifunction $\overline{\eta}_{l^*_i}$ ($\eta_{l^*_i}$) implies continuity of $\overline{\eta}_{l^*_i,1}$ ($\eta_{l^*_i,1}$ respectively).
Let us also note that if a skew product $F \in T^1_*(I)$ with a quotient map of type $\succ 2^\infty$ satisfies condition **H** (strong condition **H**) then the sequence $\{\overline{\eta}_{l^*_i}\}_{i \geq 1}$ (the sequence $\{\eta_{l^*_i}\}_{i \geq 1}$) can only contain a finitely many discontinuous functions.

We denote the subspace of $T^1_*(I)$ consisting of skew products whose quotient maps have type $\succ 2^\infty$ and satisfy strong condition **H** by $T^1_{*,1}(I)$, and the subspace of $T^1_*(I)$ consisting of skew products whose quotients have type $\succ 2^\infty$ and satisfy condition **H**, but do not satisfy strong condition **H**, by $T^1_{*,2}(I)$. We let $T^1_{*,3}(I)$ denote the subspace of $T^1_*(I)$ consisting of maps with quotients of type $\succ 2^\infty$, each of which has a sequence of suitable functions $\{\overline{\eta}_{l_i}\}_{i>0}$ containing infinitely many discontinuous functions and has a continuous Ω-function. Finally, we let $T^1_{*,4}(I)$ denote the subspace of $T^1_*(I)$ consisting of maps with quotients of type $\succ 2^\infty$, each of which has a sequence of suitable functions $\{\overline{\eta}_{l_i}\}_{i>0}$ containing infinitely many discontinuous functions and has a discontinuous Ω-function. The subspaces $T^1_{*,1}(I) - T^1_{*,4}(I)$ are pairwise disjoint.

Decomposition Theorem [17, 18]. *Each of the subspaces* $T^1_{*,i}(I)$ *for* $1 \leq i \leq 4$ *is nonempty, and their union* $\bigcup_{i=1}^{4} T^1_{*,i}(I)$ *coincides with the part of the space* $T^1_*(I)$ *consisting of skew products with quotients of type* $\succ 2^\infty$.

Definition of subspaces $T^1_{*,i}(I)$ $(1 \leq i \leq 4)$ is based on the use of special multifunctions. Therefore, these subspaces require an explicit functional description. We begin such description in this paper.

2. Properties of Skew Products with Stable as a Whole in C^1-norm Family of Fibers Maps

Let $\widetilde{T}^1_*(I)$ be the subspace of maps $F \in T^1_*(I)$ satisfying the inclusion

$$F(\partial I) \subseteq \partial I,$$

[3] For any map of a closed interval of type $\succ 2^\infty$, there exists a quasiminimal set such that the natural numbers $m_0(K(f))$, $i_0(K(f))$ and the finite subset $N_*(K(f))$ of $\mathbf{N}$ are defined by the Sharkovsky order (see [24].

[4] We assume that $n_* = 1$ if $N_*(K(f)) = \emptyset$ for any $K(f) \subset \Omega_p(f)$.

where ∂I is the boundary of I.

Definition 4 [15]. We say that *a family of fibers maps of a skew product* $F \in \widetilde{T}^1_*(I)$ *with a quotient map of type* $\succ 2^\infty$ *is stable as a whole in* C^1*-norm* if for any $\delta > 0$ there is a neighborhood $B^1_\varepsilon(F)$ of F in $\widetilde{T}^1_*(I)$ such that for any map $\Phi \in B^1_\varepsilon(F)$, where $\Phi(x,\, y) = (\varphi(x),\, \psi_x(y))$, and any l^*_i ($i \geq i^*$ for some $i^* \geq i_*$) one can find δ-close to the identity map in C^0-norm homeomorphism

$$H^{<l^*_i>} : \overline{\eta}^F_{l^*_i} \to \overline{\eta}^\Phi_{l^*_i} \qquad (H^{<l^*_i>}(x,\, y) = (h_1(x),\, h^{<l^*_i>}_{2,\,x}(y)))$$

satisfying the equality

$$h^{<l^*_i>}_{2,\,x}\,{}_{|\overline{\eta}^F_{l^*_i}(x)} \circ g_{x,\,l^*_i}\,{}_{|\overline{\eta}^F_{l^*_i}(x)}(y) = \psi_{h_1(x),\,l^*_i}\,{}_{|\overline{\eta}^\Phi_{l^*_i}(h_1(x))} \circ h^{<l^*_i>}_{2,\,x}\,{}_{|\overline{\eta}^F_{l^*_i}(x)}(y), \tag{6}$$

where $(x;\, y)$ is a point of the graph of a function $\overline{\eta}^F_{l^*_i}$ in I.

Let us note that by Definition 4 and by the choice of numbers l^*_i (see equality (5)), iterations of a skew product $F \in \widetilde{T}^1_*(I)$ with a quotient map of type $\succ 2^\infty$ and with stable as a whole in C^1-norm family of fibers maps have also stable as a whole in C^1-norm families of fibers maps.

To give the criterion of stability as a whole in C^1-norm of a family of fibers maps of a skew product $F \in \widetilde{T}^1_*(I)$ with a quotient map of type $\succ 2^\infty$ we use auxiliary skew products $F_{l^*_i}(x,\, y) = (x,\, g_{x,\,l^*_i}(y))$ and $\Phi_{l^*_i}(x,\, y) = (x,\, \psi_{x,\,l^*_i}(y))$ for corresponding iterations of F and Φ respectively. As it follows from [27] the graphs of multifunctions $\overline{\eta}^F_{l^*_i}$ and $\overline{\eta}^\Phi_{l^*_i}$ in I coincide with nonwandering sets of the maps $F_{l^*_i\,|\Omega(f)\times I_2}$ and $\Phi_{l^*_i\,|\Omega(\varphi)\times I_2}$ respectively.

Therefore, the following result is the direct corollary of Definition 4.

Theorem 1. *A family of fibers maps of a skew product $F \in \widetilde{T}^1_*(I)$ with a quotient map of type $\succ 2^\infty$ is stable as a whole in C^1-norm iff for any $\delta > 0$ there exists a neighborhood $B^1_\varepsilon(F)$ of F in $\widetilde{T}^1_*(I)$ such that for any map $\Phi \in B^1_\varepsilon(F)$ and any l^*_i ($i \geq i^*$ for $i^* \geq i_*$) one can find δ-close to the identity map in C^0-norm homeomorphism $H^{<l^*_i>} : \overline{\eta}^F_{l^*_i} \to \overline{\eta}^\Phi_{l^*_i}$ of skew products class possessing the property:*
maps $F_{l^*_i\,|\Omega(f)\times I_2}$ and $\Phi_{l^*_i\,|\Omega(\varphi)\times I_2}$ are Ω-conjugate under homeomorphism $H^{<l^*_i>}$.

Let us use Theorem 1 and the above definition of maps $F_{l^*_i}$ and $\Phi_{l^*_i}$ for skew products $F \in \widetilde{T}^1_*(I)$ and $\Phi \in B^1_\varepsilon(F)$ respectively (with quotients of type $\succ 2^\infty$ and stable as a whole in C^1-norm families of fibers maps). Then we have

$$H^{<l^*_i>}(\{x\} \times (\overline{\eta}^F_{l^*_i})(x)) = \{h_1(x)\} \times (\overline{\eta}^\Phi_{l^*_i})(h_1(x)).$$

By Definition 2 the equalities $\eta^F_{l^*_i}(x) = \Omega(g_{x,\,l^*_i})$ and $\eta^\Phi_{l^*_i}(x) = \Omega(\psi_{x,\,l^*_i})$ hold. Since $H^{<l^*_i>}$ is homeomorphism, then for any $x \in \Omega_p(f)$ the equality

$$H^{<l^*_i>}(\{x\} \times \eta^F_{l^*_i}(x)) = \{h_1(x)\} \times \eta^\Phi_{l^*_i}(h_1(x))$$

is valid. Hence, the equality $H^{<l^*_i>}(\overline{\eta}^F_{l^*_i}) = \overline{\eta}^\Phi_{l^*_i}$ is fulfilled simultaneously with the equality

$$H^{<l^*_i>}(\eta^F_{l^*_i}) = \eta^\Phi_{l^*_i}$$

and inequality

$$dist_I(\eta^F_{l^*_i},\, \eta^\Phi_{l^*_i}) < \delta,$$

where $dist_I$ is Hausdorff metric in the space 2^I of all closed subsets of the phase space I (for details, see [22, 23]).

Completing these preliminary considerations, one can prove the following statement.

Theorem 2. *Let $F \in \widetilde{T}^1_*(I)$ have a quotient map of type $\succ 2^\infty$ and have a stable as a whole in C^1-norm family of fibers maps. Then F satisfies strong condition* **H** *and has a continuous Ω-function.*

Corollary 1. *Let a skew product $F \in \widetilde{T}^1_*(I)$ have a quotient map of type $\succ 2^\infty$ and have a stable as a whole in C^1-norm family of fibers maps. Then there exists a neighborhood $U_{1,\delta'}(\Omega(f))$ in I_1 of the set $\Omega(f)$ such that for every $x \in U_{1,\delta'}(\Omega(f))$ and l_i^* for $i \geq i^*$ a map g_{x,l_i^*} is Ω-stable in C^1-norm with respect to the family of fibers maps $\{g_{x,l_i^*}\}_{x \in U_{1,\delta'}(\Omega(f))}$.*

The above Corollary 1 shows that extensions $(\eta^F_{l_i^*})^{ex}$ of auxiliary functions $\eta^F_{l_i^*}$ on a neighborhood $U_{1,\delta'}(\Omega(f))$ of the set $\Omega(f)$ are continuous for every $i \geq i^*$. Therefore, fibers maps over points of the set $U_{1,\delta'}(\Omega(f)) \setminus \Omega(f)$ do not influence on the structure of the set $\Omega(F)$.

Corollary 2. *Let a skew product $F \in \widetilde{T}^1_*(I)$ have a quotient map of type $\succ 2^\infty$ and stable as a whole in C^1-norm family of fibers maps. Then, for any different points $x',\, x'' \in Per(f) \bigcap K(f)$ with the (least) periods $m(x')$ and $m(x'')$ respectively ($K(f)$ is a locally maximal quasiminimal set of f), maps $\widetilde{g}_{x'}^{\,m/m(x')}$ and $\widetilde{g}_{x''}^{\,m/m(x'')}$ are Ω-conjugate, where m is the least common multiple of numbers $m(x')$ and $m(x'')$; the notation $\widetilde{g}_x$ for every f-periodic point x means the composition $g_{f^{m-1}(x)} \circ \ldots \circ g_x$ of fibers maps over the points of periodic orbit $\{x,\, f(x), \ldots, f^{m-1}(x)\}$ (see equality (2)).*

Corollary 3. *Let a map $F \in T^1_*(I)$ satisfy conditions of Theorem 2. Then the equality*

$$\Omega(F|_{\Omega_p^*(F)}) = \overline{\bigcup_{x \in Per_p^*(f)} \{x\} \times \Omega(\widetilde{g}_x)}$$

holds for an arbitrary, everywhere dense invariant subset $Per_p^*(f)$ of the set of all periodic points of $\Omega_p(f)$, where $\Omega_p^*(F) = \Omega_p(f) \times I_2$.

3. Criterion of C^1- Ω-stability

We begin this part of the paper giving the standard definition of C^1- Ω-stability with respect to homeomorphisms of skew products class.

Definition 5. We say that a map $F \in \widetilde{T}^1_*(I)$ is *Ω-stable in C^1-norm (in the space $\widetilde{T}^1_*(I)$)* if for any $\delta > 0$ there exists $\varepsilon > 0$ such that for any map $\Phi \in B^1_\varepsilon(F)$, $\Phi(x,\, y) = (\varphi(x),\, \psi_x(y))$, one can find δ-close in C^0-norm to the identity map homeomorphism

$$H : \Omega(F) \to \Omega(\Phi),\;\; H(x,\, y) = (h_1(x),\, h_{2,\,x}(y)),$$

such that for every point $(x;\, y) \in \Omega(F)$ the equalities hold:

$$h_1 \circ f_{|\Omega(f)}(x) = \varphi \circ h_{1|\Omega(f)}(x);$$

$$h_{2,\, f(x)|(\Omega(F))(f(x))} \circ g_{x|(\Omega(F))(x)}(y) = \psi_{h_1(x)|(\Omega(\Phi))(h_1(x))} \circ h_{2,\,x|(\Omega(F))(x)}(y).$$

Theorem 3. *A skew product $F \in \widetilde{T}^1_*(I)$ with a quotient map of type $\succ 2^\infty$ is Ω-stable in C^1-norm iff its family of fibers maps is stable as a whole in C^1-norm.*

We give here only the proof of sufficiency of the property of stability as a whole in C^1-norm of a family of fibers maps of a skew product $F \in \widetilde{T}^1_*(I)$ for its C^1- Ω-stability.

1. In fact, let a family of fibers maps of a skew product $F \in \widetilde{T}^1_*(I)$ be stable as a whole in C^1-norm. Then by Definition 4 for every $\delta > 0$ there exists a neighborhood $B^1_\varepsilon(F) \subset \widetilde{T}^1_*(I)$ such that for every map $\Phi \in B^1_\varepsilon(F)$, $\Phi(x,\, y) = (\varphi(x),\, \psi_x(y))$, and every l_i^* ($i \geq i^*$ for $i^* \geq i_*$), one can find δ-close to the identity map in C^0-norm homeomorphism $H^{<l_i^*>} : \overline{\eta}^F_{l_i^*} \to \overline{\eta}^\Phi_{l_i^*}$

$(H^{<l_i^*>}(x,\, y) = (h_1(x),\, h_{2,\,x}^{<l_i^*>}(y)))$ satisfying equalities (6).
2. Let $\Phi \in B_\varepsilon^1(F)$. We construct a homeomorphism $H : \Omega(F) \to \Omega(\Phi)$ of skew products class such that

$$H \circ F_{|\Omega(F)} = \Phi_{|\Omega(\Phi)} \circ H,$$

where first coordinate function h_1 of H is defined for φ with the use of condition of Ω-stability of f in C^1-norm (see equality (3)).
Get over construction of fibers maps for H over points of the set $Per(f)$ of f-periodic points. Let x be an arbitrary point of the set $Per(f)$ with the (least) period $m(x)$. Let a natural number $i(m(x))$, $i(m(x)) \geq i^*$, be so that $m(x)$ divides $l^*_{i(m(x))}$. Set

$$h_{2,\,x|\Omega(\widetilde{g}_x)} = h_{2,\,x}^{<l^*_{i(m(x))}>}{}_{|\eta^F_{l^*_{i(m(x))}}(x)}, \tag{7}$$

where

$$\eta^F_{l^*_{i(m(x))}}(x) = \Omega(\widetilde{g}_x^{\,l^*_{i(m(x))}/m(x)}).$$

By Corollary 1 we have

$$\Omega(\widetilde{g}_x^{\,l^*_{i(m(x))}/m(x)}) = \Omega(\widetilde{g}_x).$$

As it follows from paper [27], equality (7) and Corollary 3, a map $h_{2,\,x}(y)$ is defined on the completely invariant set $\bigcup\limits_{x \in Per(f)} \{x\} \times \Omega(\widetilde{g}_x)$, which is everywhere dense in $\Omega(F)$; moreover, $h_{2,\,x}(y)$ is one to one map on $\bigcup\limits_{x \in Per(f)} \{x\} \times \Omega(\widetilde{g}_x)$. Continuity of $h_{2,\,x}(y)$ on $\bigcup\limits_{x \in Per(f)} \{x\} \times \Omega(\widetilde{g}_x)$ follows from Theorem 2, Corollaries 2 and 3.
3. Complete a definition of $h_{2,\,x}(y)$ on $\Omega(F)$ setting

$$h_{2,\,x}(y) = \lim_{n \to +\infty} h_{2,\,x_n}(y_n) \tag{8}$$

for any point $(x,\, y) \in \Omega(F)$ and for any convergent to $(x,\, y)$ sequence $\{(x_n,\, y_n)\}_{n \geq 1} \subset \bigcup\limits_{x \in Per(f)} \{x\} \times \Omega(\widetilde{g}_x)$.
Equality (8) defines correctly second coordinate function $h_{2,\,x}(y)$ of homeomorphism H on the set $\Omega(F)$. In fact, take a point $(x,\, y) \in \Omega(F)$. Let $\{(x_n,\, y_n)\}_{n \geq 1}$ and $\{(x'_n,\, y'_n)\}_{n \geq 1}$ be any different, convergent to $(x,\, y)$ sequences of points of the set $\bigcup\limits_{x \in Per(f)} \{x\} \times \Omega(\widetilde{g}_x)$. Using these two sequences, we form a new, convergent to $(x,\, y)$ sequence $\{(\overline{x}_n,\, \overline{y}_n)\}_{n \geq 1}$, for example, setting

$$(\overline{x}_{2n-1},\, \overline{y}_{2n-1}) = (x_n,\, y_n); \quad (\overline{x}_{2n},\, \overline{y}_{2n}) = (x'_n,\, y'_n). \tag{9}$$

Then, by continuity of $h_{2,\,x}(y)$ on the set $\bigcup\limits_{x \in Per(f)} \{x\} \times \Omega(\widetilde{g}_x)$ sequence $\{h_{2,\,\overline{x}_n}(\overline{y}_n)\}_{n \geq 1}$ is fundamental and, hence, it converges. Using definition of a sequence $\{(\overline{x}_n,\, \overline{y}_n)\}_{n \geq 1}$ and equalities (8) – (9) we obtain

$$\lim_{n \to +\infty} h_{2,\,\overline{x}_n}(\overline{y}_n) = \lim_{n \to +\infty} h_{2,\,x'_n}(y'_n) = \lim_{n \to +\infty} h_{2,\,x_n}(y_n) = h_{2,\,x}(y).$$

Thus, the value of the function $h_{2,\,x}(y)$ in any point $(x,\, y) \in \Omega(F)$ does not depend on the choice of a sequence $\{(x_n,\, y_n)\}_{n \geq 1} \subset \bigcup\limits_{x \in Per(f)} \{x\} \times \Omega(\widetilde{g}_x)$ convergent to a point $(x,\, y)$.
4. Verify that the above map $H : \Omega(F) \to \Omega(\Phi)$ is homeomorphism. In fact, as it follows from its definition, H is surjection of the set $\Omega(F)$ on the set $\Omega(\Phi)$. By compactness of I the map H

Journal of Physics: Conference Series **692** (2016) 012010 doi:10.1088/1742-6596/692/1/012010

is closed. Therefore, H is mutually continuous (see [22, 23]).
Prove that H is one to one map. Suppose the contrary. Since $h_1 : \Omega(f) \to \Omega(\varphi)$ is homeomorphism, then there exist different points (x, y_1) and (x, y_2), where $x \notin Per(f)$, satisfying

$$h_{2,x}(y_1) = h_{2,x}(y_2). \tag{10}$$

Let $\{(x_n, y_n^1)\}_{n\geq 1}$ and $\{(x_n, y_n^2)\}_{n\geq 1}$ be any convergent to (x, y_1) and (x, y_2) respectively sequences of points of the set $\bigcup\limits_{x\in Per(f)} \{x\}\times\Omega(\widetilde{g}_x)$. Then by equality (10) corresponding sequence $\{H(x_n, y_n^1), H(x_n, y_n^2)\}_{n\geq 1}$ is fundamental, in addition,

$$\{H(x_n, y_n^1), H(x_n, y_n^2)\}_{n\geq 1} \subset \bigcup_{\overline{x}\in Per(\varphi)} \{\overline{x}\} \times \Omega(\widetilde{\psi_{\overline{x}}}).$$

Since restriction of H on the set $\bigcup\limits_{x\in Per(f)} \{x\} \times \Omega(\widetilde{g}_x)$ is homeomorphism and

$$H(\bigcup_{x\in Per(f)} \{x\} \times \Omega(\widetilde{g}_x)) = \bigcup_{\overline{x}\in Per(\varphi)} \{\overline{x}\} \times \Omega(\widetilde{\psi_{\overline{x}}}),$$

then a sequence $\{(x_n, y_n^1), (x_n, y_n^2)\}_{n\geq 1}$ is also fundamental. Hence, $y_1 = y_2$, and the map H is one to one. This property with mutual continuity of H means that H is homeomorphism (see [22, 23]). Hence, $F \in \widetilde{T}^1_*(I)$ is Ω-stable in C^1-norm.

Corollary 4. *Let a skew product $F \in \widetilde{T}^1_*(I)$ with a quotient map of type $\succ 2^\infty$ be Ω-stable in C^1-norm. Then $F \in \widetilde{T}^1_{*,1}(I)$, and the Ω-function of F is continuous.*

Corollary 5. *Let a skew product $F \in \widetilde{T}^1_*(I)$ satisfy conditions of Corollary 4. Then for any different periodic points x', x'' ($m(x')$ and $m(x'')$ are the (least) periods of x' and x'' respectively) of every locally maximal quasiminimal set $K(f)$ of the quotient map f, fibers maps $\widetilde{g}_{x'}^{m'/m(x')}$ and $\widetilde{g}_{x''}^{m'/m(x'')}$ are Ω-conjugate, where m' is the least common multiple of numbers $m(x')$ and $m(x")$.*

4. Ω-Stable Skew Products of Maps of an Interval Are Not Dense in $\widetilde{T}^1_{*,1}(I)$

Nongenericity of Ω-stability for C^r-diffeomorphisms ($r \geq 2$) in dimensions ≥ 3 is proved in [1] and in dimension 2 follows from [7].
In this section we prove that C^1-smooth Ω-stable skew products of maps of an interval do not form everywhere dense subset in the subspace $\widetilde{T}^1_{*,1}(I) = T^1_{*,1}(I) \bigcap \widetilde{T}^1_*(I)$.

Theorem 4. *There exists a skew product $F \in \widetilde{T}^1_{*,1}(I)$ such that some its neighborhood $B^1_\varepsilon(F)$ in the space $\widetilde{T}^1_{*,1}(I)$ does not contain Ω-stable skew products of maps of an interval.*

Proof. 1. In fact, let the skew product $F(x, y) = (f(x), \lambda(x)y(1-y))$ be defined on the unit square $[0, 1]^2$. Here the quotient map is given by the equality

$$f(x) = \begin{cases} \widetilde{h}(x), & \text{if } x \in [0, \frac{1}{4}); \\ 9(\frac{1}{4} - x)(x - \frac{3}{4}) + \frac{1}{4}, & \text{if } x \in [\frac{1}{4}, \frac{3}{4}); \\ \widetilde{h}(1-x), & \text{if } x \in [\frac{3}{4}, 1]. \end{cases} \tag{11}$$

We use the function $\widetilde{h}(x) = \alpha \sin^2 \pi\beta x$ for $x \in [0, 1/4]$; moreover, $\alpha > 0$ and $\beta \in (4, 6)$ are defined uniquely from the system of equations

$$\begin{cases} \alpha \sin^2 \pi\beta/4 = 1/4; \\ 2\alpha\pi\beta \sin \pi\beta/4 \cos \pi\beta/4 = 9/2. \end{cases}$$

This system is equivalent to the condition of C^1-smoothness of f in the point $x = 1/4$. Then $f \in C^1_\omega(I_1)$, and $\Omega(f) = \{0\} \bigcup K(f)$, where $K(f)$ is the unique locally maximal quasiminimal set of the quotient map f, $K(f) = \Omega_p(f) \subset [\frac{1}{4}, \frac{3}{4}]$, $\tau(f_{|K(f)}) = \mathbf{N}$.
Note that the quotient f has two fixed points $x_1 = 1/4$, $x_2 = 23/36$ on the closed interval $[1/4; 3/4]$.
Let $\lambda = \lambda(x)$ be C^1-smooth function $[0, 1]^2$ given by formula

$$\lambda(x) = \begin{cases} \frac{9\pi}{7}(\lambda^* - 2)(x - \frac{1}{4}) + 2, & \text{if } x \in [0, \frac{1}{4}); \\ (\lambda^* - 2)\sin\frac{9\pi}{7}(x - \frac{1}{4}) + 2, & \text{if } x \in [\frac{1}{4}, \frac{23}{36}); \\ \lambda^*, & \text{if } x \in [\frac{23}{36}, 1], \end{cases} \tag{12}$$

where the number λ^* is such that the map $\overline{y} = \lambda^* y(1-y)$ has type 2^∞ (i. e. f contains periodic points of periods $1, 2, 2^2, \ldots, 2^n, \ldots$ and does not contain periodic points of other periods), $\lambda^* \approx 3,569$.
Call attention on the properties of fibers maps over f-fixed points x_1 and x_2.
(i_F) The fiber map $\widetilde{g}_{x_1}$ is defined by the equality $\widetilde{g}_{x_1} = 2y(1-y)$. Hence, $\widetilde{g}_{x_1}$ is Ω-stable in the space $C^1_{\partial_2}(I_2)$ Morse-Smale endomorphism with two hyperbolic fixed points ($\widetilde{g}_{x_1}$ does not contain other periodic points). It means that there is a ε_1-neighborhood $B^1_{2,\varepsilon_1}(\widetilde{g}_{x_1})$ of the map $\widetilde{g}_{x_1}$ in the space $C^1_{\partial_2}(I_2)$ such that every map from this neighborhood is Ω-conjugate with $\widetilde{g}_{x_1}$.
(ii_F) The fiber map $\widetilde{g}_{x_2}$ is defined by the equality $\widetilde{g}_{x_2} = \lambda^* y(1-y)$. In this case there exists strictly decreasing sequence of positive numbers $\{\varepsilon_r\}_{r\geq 2}$, $\lim\limits_{r\to+\infty} \varepsilon_r = 0$, such that every fiber map $g_x \in B^1_{2,\varepsilon_r}(\widetilde{g}_{x_2})$ for $x \in (1/4; 23/36)$ (x belongs also to some left-side neighborhood of the point x_2 defined for ε_r) contains periodic points with the (least) periods $\{1, 2, \ldots, 2^{\mu_r}\}_{0<\mu_r<\infty}$. Here $B^1_{2,\varepsilon_r}(\widetilde{g}_{x_2})$ is a ε_r-neighborhood of the map $\widetilde{g}_{x_2}$ in the space $C^1_{\partial_2}(I_2)$; a sequence $\{\mu_r\}_{r\geq 2}$ strictly increases[5] [28].
2. Since $\tau(f_{|K(f)}) = \mathbf{N}$ then $l^*_i = i$ $(i \geq 1)$. Using properties of period doubling bifurcations in one-dimensional maps one can prove that the constructed above skew product F satisfies strong condition $\mathbf{H}$. This property (with equalities (11), (12) and formulas for fibers maps) means that $F \in \widetilde{T}^1_{*,1}(I)$.
3. Point out a neighborhood $B^1_\varepsilon(F)$ in the space $\widetilde{T}^1_{*,1}(I)$ which does not contain Ω-stable skew products. In fact, let a natural number $r \geq 2$ be such that each map from the neighborhood $B^1_{2,\varepsilon_r}(\widetilde{g}_{x_2})$ contains periodic points with (least) periods $\{1, 2, \ldots, 2^{\mu_r}\}$ for $3 \leq \mu_r < \infty$. Let ε_1-neighborhood $B^1_{2,\varepsilon_1}(\widetilde{g}_{x_1})$ of the map $\widetilde{g}_{x_1}$ in $C^1_\omega(I_2)$ be chosen accordingly item 1 of the proof such that property (i_F) is fulfilled.
Set $\overline{\varepsilon} = \min\{\varepsilon_1, \varepsilon_r\}$. Use uniform continuity of C^1-presentation ρ_1 (see [29]) on the compact $K(f)$. For number $\overline{\varepsilon} > 0$ we find $0 < \delta_1 < 1/16$ such that for any x', $x'' \in K(f)$ satisfying the inequality $|x' - x''| < \delta_1$, we have $g_{x''} \in B^1_{2,\overline{\varepsilon}}(g_{x'})$. Here $\rho_1(x') = g_{x'}$, $\rho_1(x'') = g_{x''}$.
Since $f \in C^1_\omega(I)$ then for $\delta_1 > 0$, there exists $\overline{\varepsilon}'$-neighborhood $B^1_{1,\overline{\varepsilon}'}(f)$ of the quotient f in the space $C^1_\omega(I)$ consisting of maps each of them is Ω-conjugate with f over homeomorphism δ_1-close to the identity map. Point out that δ_1-neighborhoods of f-fixed points $U_{1,\delta_1}(0)$, $U_{1,\delta_1}(1/4)$ $U_{1,\delta_1}(23/36)$ for $0 < \delta_1 < 1/16$ are pairwise disjoint, and each of them contains unique fixed point of any map from $B^1_{1,\overline{\varepsilon}'}(f)$.
Set $\varepsilon = \min\{\overline{\varepsilon}, \overline{\varepsilon}'\}$. Prove that a neighborhood $B^1_\varepsilon(F)$ of the skew product $F \in \widetilde{T}^1_{*,1}(I)$ does

[5] Maps $\widetilde{g}^n_{x_1}$ and $\widetilde{g}^n_{x_2}$ for all $n \geq 1$ possess properties analogous properties (i_F) – (ii_F). But a sequence $\{\mu_r(n)\}_{r(n)\geq 1}$ for $n > 1$ can be nonmonotone. One can explain it using definition of fibers maps $g_{x,n}$ as compositions of logistic maps with different values of function $\lambda(x)$ in the points $x, f(x), \ldots f^{n-1}(x)$ (see equality (2)).

not contain Ω-stable maps. Suppose the contrary. Then there exists Ω-stable skew product $\Phi \in \widetilde{T}^1_{*,1}(I)$ (see Corollary 4) such that $\Phi \in B^1_{\varepsilon}(F)$, where $\Phi(x, y) = (\varphi(x), \psi_x(y))$. A map Φ possesses the following properties:
(i_Φ) $\varphi \in B^1_{1,\overline{\varepsilon}'}(f)$;
(ii_Φ) φ has unique locally maximal quasiminimal set $K(\varphi)$ containing two fixed points $x_1(\varphi) \in U_{1,\delta_1}(1/4)$ and $x_2(\varphi) \in U_{1,\delta_1}(23/36)$;
(iii_Φ) $\widetilde{\psi}_{x_1(\varphi)} \in B^1_{2,\varepsilon_1}(\widetilde{g}_{x_1})$; (iv_Φ) $\widetilde{\psi}_{x_2(\varphi)} \in B^1_{2,\varepsilon_r}(\widetilde{g}_{x_2})$.
By properties (i_F) and (iii_Φ) the map $\widetilde{\psi}_{x_1(\varphi)}$ is Morse-Smale endomorphism with two fixed points. At the same time, by properties (ii_F) and (iv_Φ) the map $\widetilde{\psi}_{x_2(\varphi)}$ contains periodic orbits with (least) periods $\{1, 2, \ldots, 2^{\mu_r}\}$ for $3 \le \mu_r < \infty$. Therefore, there are no iterations of maps $\widetilde{\psi}_{x_1(\varphi)}$ and $\widetilde{\psi}_{x_2(\varphi)}$ which are Ω-conjugate.
Contradiction with Corollary 5 is obtained. Theorem 4 is proved.

Acknowledgments

Results of Sec. 1, 2 are supported by Ministry of Science and Education of Russia, grant No 14–10. Results of Sec. 3, 4 are supported by Russian Science Foundation, grant No 16-11-10036. Investigations of Sec. 3, 4 are executed by the Author in Moscow State University.

References

[1] Abraham R and Smale S 1970 *Global Analysis* **14** 5–8
[2] Anosov D 1985 *Proceedings of the Steklov Institute of Mathematics* **169** 61-95
[3] Hayashi Sh 1997 *Ann. Math.* **145** 81–137
[4] Hayashi Sh 1999 *Ann. Math.* **150** 353–6
[5] Mañé R 1988 *Publs. Math. IHES* **66** 139–59
[6] Mañé R 1988 *Publs. Math. IHES* **66** 161–210
[7] Newhouse Sh 1970 *Global Analysis* **14** 191–202
[8] Palis J 1988 *Publs. Math. IHES 66* 211–5
[9] Smale S 1967 *Bull. Amer. Math. Soc.* **73** 747-817
[10] Jakobson M 1971 *Mathematics of the USSR – Sbornik* **14** 161–85
[11] Ikeda H 1996 *Trans. Amer. Math. Soc.* **348** 2183-200
[12] Przytycki F 1977 *Stud. Math.* **60** 61-77
[13] Efremova L 1999 *Itogi Nauki Tekh. Ser. Sovrem. Mat. Prilozh. Temat. Obz., Vseross. Inst. Nauchn. i Tekhn. Inform. (VINITI)* **67** 129–60
[14] Efremova L 2001 *Nonlinear Analysis* **47** 5297–308
[15] Efremova L 2002 *Progress in Nonlinear Science* **1** 219–24.
[16] Efremova L 2002 *Proceedings of the Steklov Institute of Mathematics* **236** 157-63
[17] Efremova L 2010 *Theor. and Math. Physics* **164** 1208-14
[18] Efremova L 2013 *Sb. Math.* **204** 1598–623
[19] Katok A B and Hasselblatt B 1995 *Introduction to the Modern Theory of Dynamical Systems. Encyclopedia Math. Appl.* (Cambridge: Cambridge Univ. Press)
[20] de Melo W and van Strien S 1993 *One-dimensional Dynamics. A Series of Modern Surveys in Mathematics* (Berlin - Heidelberg: Springer-Verlag)
[21] Nemytskii V and Stepanov V 1947 *Qualitative theory of differential equations* (Moscow - Leningrad: OGIZ)
[22] Kuratowski K 1966 *Topology* **1** (New York-London: Academic Press)
[23] Kuratowski K 1968 *Topology* **2** (New York-London: Academic Press)
[24] Šarkovs'kiĭ O 1965 *Ukrain. Mat. Z.* **17** 104-11
[25] Efremova L 1981 *Periodic motions of discrete semidynamical systems. PhD diss.* (Gorky: Gorky State University)
[26] Alseada L and Llibre J 1985 *Proc. Amer. Math. Soc.* **93** 133–8
[27] Efremova L 2014 *Springer Proceedings in Math. and Statistics* ed Gracio C, Fournier-Prunaret D, Ueta T and Nishio Y (New York: Springer Science) vol **57** 39–58
[28] Misiurewicz M 1981 *Publs. Math. IHES* **53** 5–16
[29] Nitecki Z 1971 *Differetiable Dynamics* (Cambridge, Massachusetts and London: MIT Press)

Robust PRNG based on homogeneously distributed chaotic dynamics

Oleg Garasym[1], René Lozi[2] and Ina Taralova[1]

[1]Ecole Centrale de Nantes, France; [2]Laboratoire J. A. Dieudonné, UMR CNRS 7351 Nice Sophia-Antipolis, France

E-mail: oleg.garasym@irccyn.ec-nantes.fr, rlozi@unice.fr and ina.taralova@irccyn.ec-nantes.fr

Abstract. This paper is devoted to the design of new chaotic Pseudo Random Number Generator (CPRNG). Exploring several topologies of network of 1-D coupled chaotic mapping, we focus first on two dimensional networks. Two topologically coupled maps are studied: TTL^{RC} non-alternate, and TTL^{SC} alternate. The primary idea of the novel maps has been based on an original coupling of the tent and logistic maps to achieve excellent random properties and homogeneous /uniform/ density in the phase plane, thus guaranteeing maximum security when used for chaos base cryptography. In this aim two new nonlinear CPRNG: $MTTL_2^{SC}$ and $NTTL_2$ are proposed. The maps successfully passed numerous statistical, graphical and numerical tests, due to proposed ring coupling and injection mechanisms.

1. Introduction

tremendous development of new IT technologies, e-banking, e-purchasing, etc. nowadays increases incessantly the needs for new and more secure cryptosystems. The latter are used for information encryption, pushing forward the demand for more efficient and secure pseudo-random number generators [1]. At the same time, chaotic maps show up as perfect candidates able to generate independent and secure pseudo-random sequences (used as information carriers or directly involved in the process of encryption/decryption). However, the majority of well-known chaotic maps are not naturally suitable for encryption [2] and most of them don't exhibit even satisfactory properties for encryption. Generally, it is known [3] that true randomness is only obtained using a nondeterministic physical process like radioactivity, pseudo-randomness corresponds to a mathematical algorithm satisfying statistical properties and chaos is limited to dynamical systems. Some authors [4] refer flawly to true or truly random numbers while they are involving only chaotic numbers in their articles. True randomness is never reproducible, which forbids the use of true random numbers in cryptographic algorithms which need synchronisation. Electronic circuits cannot be used because small variations in temperature change the characteristic function which models the behavior, therefore it could not be exactly reproduced. Therefore, we deal with pseudo-random number generator (PRNG)and in this aim we propose the original idea to couple tent and logistic map, and to add an injection mechanism to keep bounded the escaping orbits.

In 1973, sir Robert May, a famous biologist introduced the nonlinear, discrete time dynamical system called logistic equation:

Journal of Physics: Conference Series **692** (2016) 012011 doi:10.1088/1742-6596/692/1/012011

$$x_{n+1} = rx_n(1 - x_n) \tag{1}$$

as a model for the fluctuations in the population of fruit flies in a closed container with constant food [5]. Since that early time this logistic equation has been extensively studied especially by May [6], and Feigenbaum [7] under the equivalent form:

$$x_{n+1} = f_\mu(x_n) \tag{2}$$

where

$$f_\mu(x_n) \equiv L_\mu(x) = 1 - \mu x^2 \tag{3}$$

Another often studied discrete dynamical system is defined by the symmetric tent map:

$$f_\mu(x_n) \equiv T_\mu(x_n) = 1 - \mu|x| \tag{4}$$

In both cases, μ is a control parameter that has impact to chaotic degree, and those mappings are sending the one-dimensional interval $[-1, 1]$ into itself.

Those two maps have also been fully explored with the hope of generating pseudo-random numbers [8]. However the collapsing of iterates of dynamical systems or at least the existence of very short periodic orbits, their non constant invariant measure, and the easily recognized shape of the function in the phase space should lead to avoid the use of such one-dimensional map (logistic, baker, or tent, etc.) or two dimensional map (Hénon, standard or Belykh, etc.) as a pseudo-random number generator (see [9] for a survey). However, the very simple implementation in computer program of chaotic dynamical systems led some authors to use them as a base of cryptosystem [10]. They are topologically conjugate, that means they have similar topological properties (distribution, chaoticity, etc.) however due to the structure of number in computer realization their numerical behavior differs drastically. Therefore the original idea here is to combine features of tent (T_μ) and logistic (L_μ) maps to achieve new map with improved properties, trough combination of several network topologies. In this paper we propose new ideas of tent and logistic maps coupling, based on the analogy between mathematical circuits and electrical circuits [11].

2. Exploring topologies of network of coupled chaotic maps

Ring and auto-coupling of chaotic maps (or circuits) enables to combine the individual circuit's dynamics, and therefore, to obtain more complex dynamic behavior. For instance, different ways of coupling several Chua's circuits gives rise to hyperchaos [11]. It should be emphasized that these representations (Fig. 1a) are also a perfect tool to investigate the topology of the map.

Looking at the equations we can inverse the shape of the graph of the tent map T on the step of logistic map L. Thus, our proposition has the form:

$$f_\mu(x) \equiv TL_\mu(x) = \mu|x| - \mu x^2 = \mu(|x| - x^2) \tag{5}$$

Recall that both logistic and tent maps have never been used in cryptography because they have weak security (collapsing effect) [12, 13] if applied alone. Thus, systems are often used in modified form to construct PRNG. The Lozi system [14] provides method to increase randomness properties of the tent map over its coupling. In another way, we propose to couple T_μ map over combination with TL_μ map (5). When used in more than one dimension, TL_μ map can be considered as a two dimensional map:

$$TL_\mu(x^{(1)}, x^{(2)}) = \mu(|x^{(1)}| - (x^{(2)})^2) \tag{6}$$

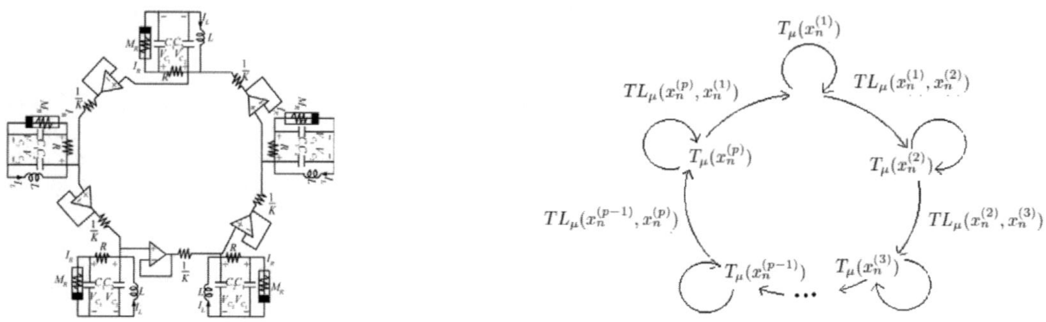

a) b)

Figure 1. a) Circuits of ultra-weak coupling of chaotic Chua's circuits b) Auto and ring-coupling between states of the M_p)

Hence it is possible to define a mapping M_p from $[-1,1]^p \to [-1,1]^p$

$$M_p \begin{pmatrix} x_n^{(1)} \\ x_n^{(2)} \\ \vdots \\ x_n^{(p)} \end{pmatrix} = \begin{pmatrix} x_{n+1}^{(1)} \\ x_{n+1}^{(2)} \\ \vdots \\ x_{n+1}^{(p)} \end{pmatrix} = \begin{cases} T_\mu(x_n^{(1)}) + TL_\mu(x_n^{(1)}, x_n^{(2)}) \\ T_\mu(x_n^{(2)}) + TL_\mu(x_n^{(2)}, x_n^{(3)}) \\ \vdots \\ T_\mu(x_n^{(p)}) + TL_\mu(x_n^{(p)}, x_n^{(1)}) \end{cases} \tag{7}$$

Note that the system dynamics is unstable and trajectories quickly spread out. Therefore, to solve the problem of holding dynamics in the bound $[-1,1]^p$ the following injection mechanism has to be used:

$$\begin{array}{c} \text{if } x_{n+1}^{(i)} < -1 \\ \text{then add } 2 \\ \text{if } x_{n+1}^{(i)} > 1 \\ \text{then substract } 2 \end{array} \tag{8}$$

in this case for $1 \le i \le p$, points come back from $[-3,3]^p$ to $[-1,1]^p$.

Auto and ring-coupling between states (Fig.1b) of the map and injection mechanism influence the system dynamics making its dynamics complex enough for our application purposes.

Used in conjunction with T_μ the TL_μ function allows to establish mutual influence between system states. The function is attractive because it performs contraction and stretching distance between states improving chaotic distribution. Thus, TL_μ function is a powerful tool to change dynamics.

The coupling of the simple states has excellent effect on chaos achieving, because:

- Simple states interact with global system dynamics, being a part of it.
- The states interaction has a global effect.

Hence, if we use TL_μ to make impact on the dynamics of simple maps, then excellent effect on chaoticity and randomness could be achieved. The proposed function improve the complexity of a simple map.

Note that the system (7) can be seen in the scope of a general point of view, introducing constants k^i which generalize considered topologies. It is called non-alternate if $k^i = +1$, or $k^i = -1$, $1 \le i \le p$, or alternate if $k^i = (-1)^i$, $1 \le i \le p$. It can be a mix of alternate and non-alternate if $k^i = +1$ or -1 randomly. As well it has been already shown that the coupling could improve the performances of well known chaotic attractors (Chua, Lorenz, Rossler, etc.) for application purposes [11].

$$M_p \begin{pmatrix} x_n^{(1)} \\ x_n^{(2)} \\ \vdots \\ x_n^{(p)} \end{pmatrix} = \begin{pmatrix} x_{n+1}^{(1)} \\ x_{n+1}^{(2)} \\ \vdots \\ x_{n+1}^{(p)} \end{pmatrix} = \begin{cases} T_\mu(x_n^{(1)}) + k^1 \times TL_\mu(x_n^{(1)}, x_n^{(2)}) \\ T_\mu(x_n^{(2)}) + k^2 \times TL_\mu(x_n^{(2)}, x_n^{(3)}) \\ \vdots \\ T_\mu(x_n^{(p)}) + k^p \times TL_\mu(x_n^{(p)}, x_n^{(1)}) \end{cases} \tag{9}$$

In this paper we will discuss only systems exhibiting the best properties for CPRNG. Therefore, we will consider only two 2-D systems: $TTL_\mu^{RC}(x_n^{(2)}, x_n^{(1)})$ **non-alternate**:

$$TTL_\mu^{RC} : \begin{cases} x_{n+1}^{(1)} = 1 - \mu|x_n^{(1)}| + \mu(|x_n^{(2)}| - (x_n^{(1)})^2) \\ x_{n+1}^{(2)} = 1 - \mu|x_n^{(2)}| + \mu(|x_n^{(1)}| - (x_n^{(2)})^2) \end{cases} \tag{10}$$

and $TTL_\mu^{SC}(x_n^{(1)}, x_n^{(2)})$ **alternate**:

$$TTL_\mu^{SC} : \begin{cases} x_{n+1}^{(1)} = 1 - \mu|x_n^{(1)}| - \mu(|x_n^{(1)}| - (x_n^{(2)})^2) \\ x_{n+1}^{(2)} = 1 - \mu|x_n^{(2)}| + \mu(|x_n^{(1)}| - (x_n^{(2)})^2) \end{cases} \tag{11}$$

Here RC stand for ring-coupling and SC for standard coupling.

3. Randomness study of the new maps TTL_μ^{RC} and TTL_μ^{SC}

We are now assessing the randomness of both selected maps. The associated dynamical system is considered to be random and could be applied to cryptosystems if the chaotic generator meets the requirements 1-8 on Fig.2. If one of the criterion is not satisfied, the behavior is less random than expected.

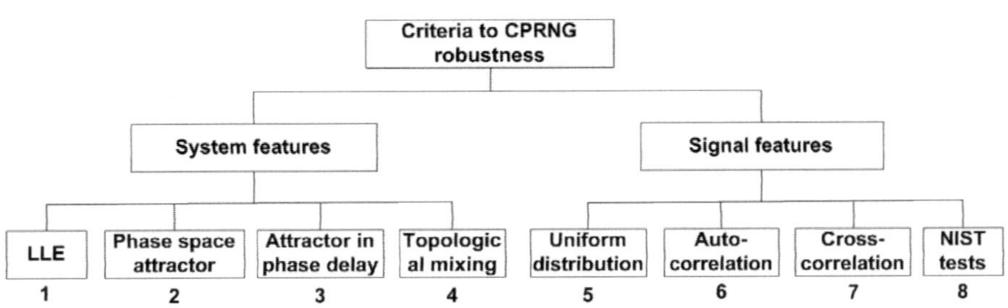

Figure 2. The main criteria for PRNG robustness

As it has been summarized in the scheme (Fig.2) a generator could be taken into consideration for cryptography application if and only if each criterion is satisfied. In the limited extend of this article we give a sketch of the assessment of some criteria such as criteria 1 (Fig. 4), 2 and 5 (Fig. 5, 7), 3 (Fig. 9), 8 (Fig. 12), for criteria 4, 6, 7 one can refer to [15].

Chaotic map behavior primarily depends on the initial guess x_0 and "control" parameter μ. However, the dependence versus the initial guess, x_0 has less importance when the global phase portrait is scrutinized. Thus, to study the dependency of parameter μ a bifurcation diagram is an appropriate tool. To create the diagram for the new map, a particular initial value of x_0 is randomly selected, and the map is iterated for a given μ. A certain number of firstly generated points is cut off to remove the transient part of the iterated points, and the following points are plotted. Afterwards, the process is repeated incrementing slightly μ.

To plot the bifurcation diagram for the 2-D systems TTL_{μ}^{RC} non-alternate (Fig. 3.a) and TTL_{μ}^{SC} alternate (Fig. 3.b), 10,000 iterations have been generated for each initial value and the first 1,000 points have been cut off as transient. Thus, 9,000 points are plotted for each μ parameter. The graphs are the same for $x^{(1)}$ and $x^{(2)}$.

a) b)

Figure 3. Bifurcation diagram of 2-D new maps a) TTL_{μ}^{RC} non-alternate (10) b) TTL_{μ}^{SC} alternate (11)

For both graphs starting from $\mu = 0$ to $\mu = 0.25$, we can observe a period 1 (*i.e.* a fixed point). Then the steady-state response undergoes a so-called pitchfork bifurcation to period 2. Following bifurcation undergoes multiple periods. At higher μ values, the behavior is generally chaotic. However, for TTL_{μ}^{RC} near $\mu = 1.1$ (Fig. 3.a) periodic windows appear. The subsequent intervals show perfect chaotic dynamics.

The Lyapunov exponent (LE) is a measure of the system sensitivity to initial conditions. The function of Lyapunov exponent λ is the characteristic of chaotic behavior in nonlinear maps. If the largest LE $\lambda > 0$ the system exhibits chaotic behavior.

Let us observe the graphics of Lyapunov exponent for TTL_{μ}^{RC} non-alternate (Fig. 4.a) and TTL_{μ}^{SC} alternate (Fig. 4.b) maps. For the plotting $10,000$ iterations were taken into account for every value of μ. The μ parameter is selected from 0.5 to 2. The list of points formed with μ is described on the horizontal coordinate and the measure λ is on the vertical coordinate.

a) b)

Figure 4. Function of the largest Lyapunov exponent for 2-D new maps a) TTL_{μ}^{RC} non-alternate map (10) b) TTL_{μ}^{SC} alternate map (11)

Graphs of the largest Lyapunov exponent are in exact agreement with those of the bifurcations. The measure λ is positive indicating chaotic dynamics which increases showing the strongest chaos at $\mu = 2$.

The study demonstrates that TTL_{μ}^{RC} non-alternate (Fig. 4.a) and TTL_{μ}^{SC} alternate (Fig. 4.b) maps exhibit the best chaotic behavior characteristics when $\mu = 2$, therefore we will continue our study fixing the parameter to this value. On the graphs for any given initial point x_0

trajectories will look like chaotic. Hence, we can study an attractor in phase space and phase delay.

The quality of the entire cryptosystem [15] mostly depends on PRNG and one of the most important things for robust PRNG is uniform distribution of generated values in the space (Criterion 5, Fig. 2). An approximated invariant measure gives the best description of probability. Thus, the invariant measure [16] is used for precise study of the points distribution. Using the approximate density function the best picture of points density can be achieved. The graph of the function demonstrates distribution comparison between regions. The size of each of the boxes is measured by *step*. In other words the plane is divided $boxes[i,j]$ with square area $step^2$, after counting the points entering into the box $box[i,j]$.

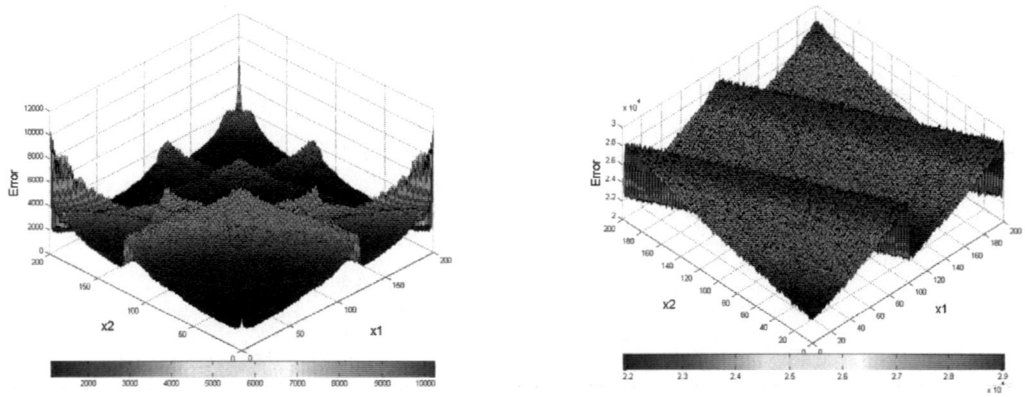

a) b)

Figure 5. Approximate density function, where $step = 0.01$, 10^9 points are generated a) TTL_2^{RC} non-alternate map b) TTL_2^{SC} alternate map

For the approximation function the pattern was divided into 200 boxes or $step = 0.01$, 10^9 points were generated. Note that those values are the maximum possible used to calculate with a laptop computers. The graphs (Figs. 5a and 5b) of the detailed points distribution demonstrates that both systems do not have excellent distribution in phase space. Note: due to the squared nonlinearity of logistic map it is impossible to define explicitly a transition matrix corresponding to a Markov chain of TTL_2^{SC}, therefore it is only possible to compute numerically its invariant measure.

Good results are demonstrated with two different kinds of coupling, simple and ring-coupling in dimension 2, thus increasing the complexity of the system. However as those results are not completely satisfactory, an improved geometry of coupling is introduced allowing us to describe a new 2-D Chaotic Pseudo Random Number Generator (CPRNG). It was noticed that some parts of the graph are perfectly joined, giving us an idea to improve the points density using some correction in the equations.

4. Two new 2-D chaotic PRNG

Considering the results of section 3 it seems possible to improve the randomness of the 2-D topology. First, let us rewrite the mapping TTL_μ^{SC} alternate (11) where $\mu = 2$ as follows:

$$TTL_2^{SC}(x_n^{(1)}, x_n^{(2)}) = \begin{cases} x_{n+1}^{(1)} = 1 + 2(x_n^{(2)})^2 - 4|x_n^{(1)}| \\ x_{n+1}^{(2)} = 1 - 2(x_n^{(2)})^2 + 2(|x_n^{(1)}| - |x_n^{(2)}|) \end{cases} \quad (12)$$

The first problem is that top green coloured region occurs after injection is applied. Thus, we develop the system (15) in such a way that green coloured region "stays" in such position without injection mechanism. Secondly, we need to reduce the width of the region. Obviously, it is possible to achieve this need by reducing the impact of the state x^1, with the new following map:

$$MTTL_2^{SC}(x_n^{(1)}, x_n^{(2)}) = \begin{cases} x_{n+1}^{(1)} = 1 + 2(x_n^{(2)})^2 - 2|x_n^{(1)}| \\ x_{n+1}^{(2)} = 1 - 2(x_n^{(2)})^2 + 2(|x_n^{(1)}| - |x_n^{(2)}|) \end{cases} \tag{13}$$

with the injection mechanism (8) used as well, but restricted to 3 phases:

$$\begin{array}{c} if\ x_{n+1}^{(1)} > 1\ then\ substract\ 2 \\ if\ x_{n+1}^{(2)} < -1\ then\ add\ 2 \\ if\ x_{n+1}^{(2)} > 1\ then\ substract\ 2 \end{array} \tag{14}$$

The results of the modifications are demonstrated on Figs. 6, 7.a and 7.b. The injection mechanism in 3 phases (Fig. 6) matches the regions in an excellent way. The techniques used, greatly improve the points density in the phase space (Figs. 7).

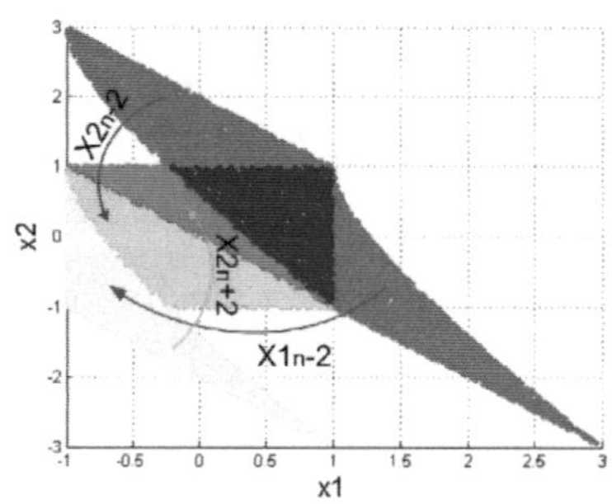

Figure 6. Injection mechanism (14) of $MTTL_2^{SC}$ alternate map

The numerical results of the errors distributions (Fig. 8) shows excellent distribution till 10^9 points which is limited by the classical computer power. Moreover, the largest Lyapunov exponent is equal to 0.5905 indicating strong chaotic behavior.

The graph (Fig. 8) shows straight error reducing that proves uniform points distribution when the number of iterates increases.

The points distribution of the attractor in phase delay is quite good as well (Fig. 9), where 10^9 points have been generated. In Fig. (9.b) tent distribution can be recognized for $x^{(2)}$ variable but for encryption we need only output of one state (in our case $x^{(1)}$). Both states make strong impact on each other and for the global dynamics, reaching significant points distribution on the torus and chaoticity.

We introduce now another structurally simple 2-D map using another topology that exhibits homogeneous dynamics distribution in phase space and phase delay space for both states. The map is described as follows:

$$NTTL_2^{SC}(x_n^{(1)}, x_n^{(2)}) = \begin{cases} x_{n+1}^{(1)} = 1 - 2|x_n^{(2)}| \\ x_{n+1}^{(2)} = 1 - 2(x_n^{(2)})^2 - 2(|x_n^{(2)}| - |x_n^{(1)}|) \end{cases} \tag{15}$$

applying injection mechanism (Fig. 10) to hold dynamics in $[-1, 1]^2$:

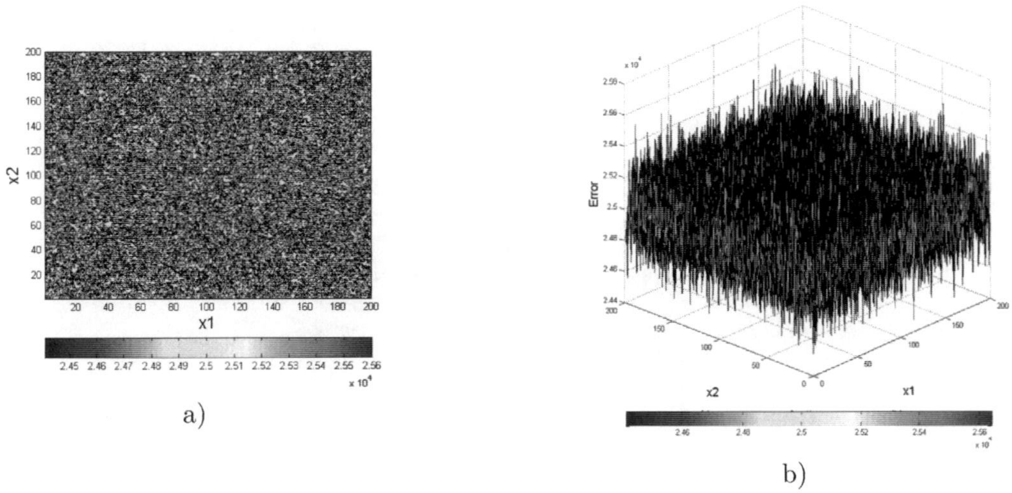

Figure 7. Approximate density function of $MTTL_2^{SC}$ alternate map, where $step = 0.01$, 10^9 points are generated a) Boxes method b) 3D

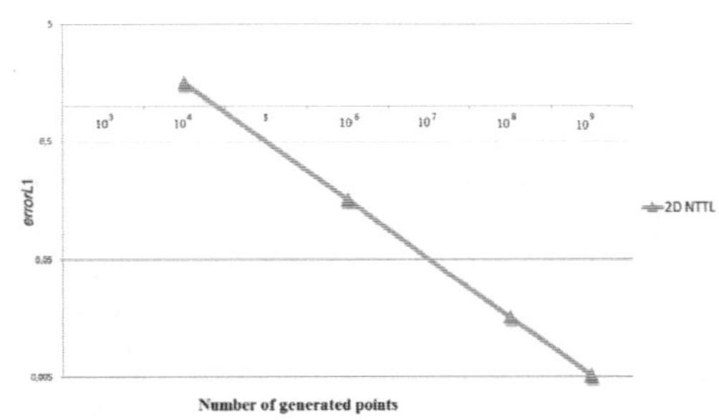

Figure 8. Approximate distribution errors, for the system (13)

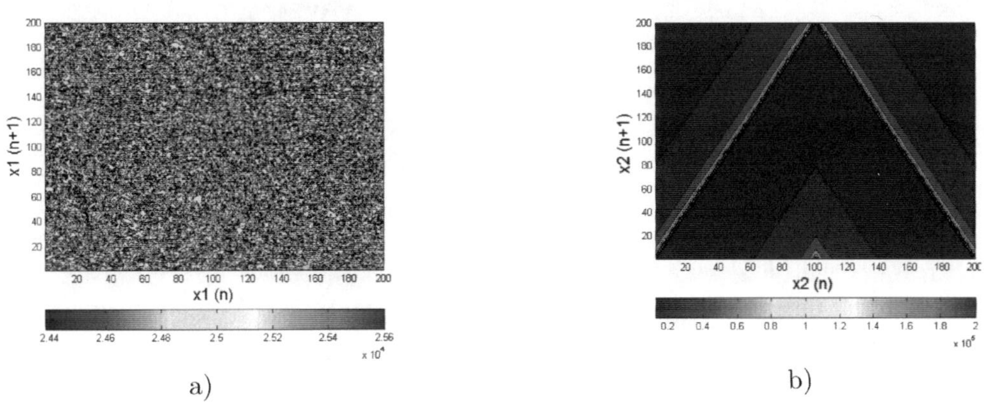

Figure 9. a) Attractor in the phase delay, 10^9 points are generated, for the system (13) a) $(x_n^{(1)}, x_{n+1}^{(1)})$ b) $(x_n^{(2)}, x_{n+1}^{(2)})$

$$\begin{array}{l} if\ x^{(2)}_{n+1} < -1 \text{ then add } 2 \\ if\ x^{(2)}_{n+1} > 1 \text{ then substract } 2 \end{array} \quad (16)$$

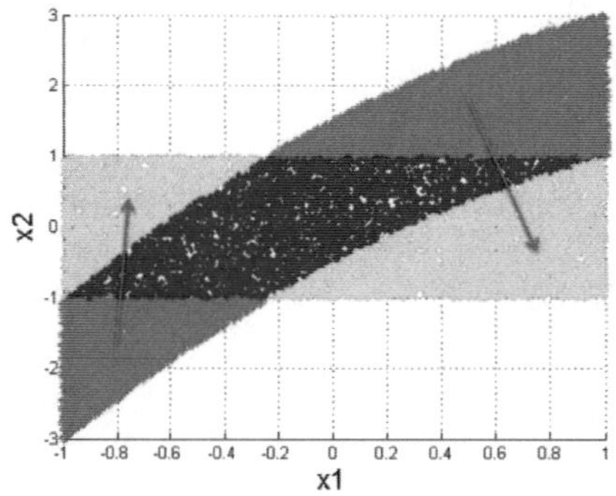

Figure 10. Injection mechanism (14) of $NTTL_2$ alternate map

The $NTTL_2$ exhibits excellent density in phase delay for both states x^1 and x^2 (Fig. 11) and successfully passed all criteria (Fig. 2), being very promising for real application.

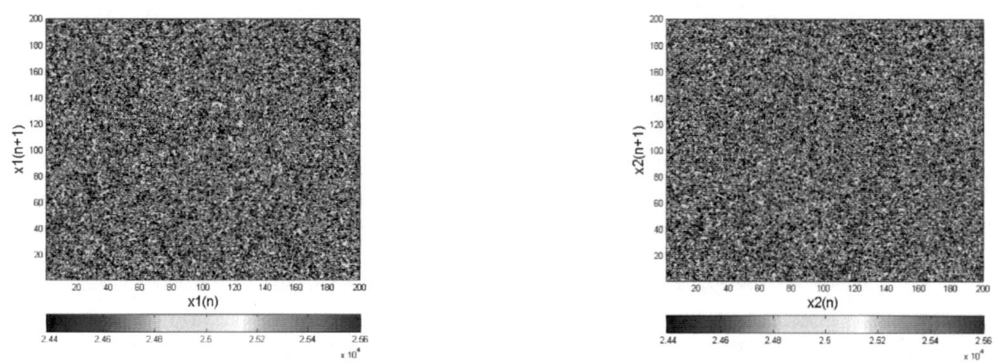

a) b)

Figure 11. $NTTL_2$ density in phase delay **a)** $(x^{(1)}_n, x^{(1)}_{n+1})$ **b)** $(x^{(2)}_n, x^{(2)}_{n+1})$

The $NTTL_2$ map exhibits complex dynamics capable to refuse statistical attacks since it successfully passed NIST tests (Fig. 12).

The future work will be devoted to the investigation of topologies in 3 dimensional space where the complexity of dynamical phenomena is expected to exhibit even better performances, though being more intricate [17]

5. Conclusion

In this paper we have proposed the original idea to couple two well-known chaotic maps (tent and logistic one), which considered separately - do not exhibit the required features for encryption purposes because they have weak security (collapsing effect) when applied alone. The new coupling changed qualitatively the overall system behavior, because the maps used with injection mechanism and coupling between states increased their complexity.

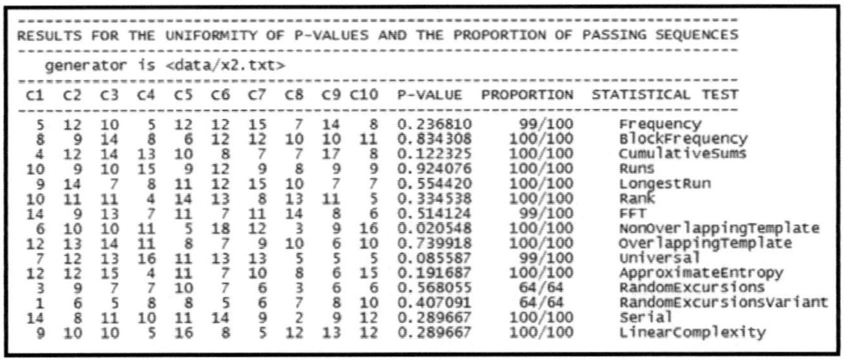
RESULTS FOR THE UNIFORMITY OF P-VALUES AND THE PROPORTION OF PASSING SEQUENCES

generator is <data/x2.txt>

C1	C2	C3	C4	C5	C6	C7	C8	C9	C10	P-VALUE	PROPORTION	STATISTICAL TEST
5	12	10	5	12	12	15	7	14	8	0.236810	99/100	Frequency
8	9	14	8	6	12	12	10	10	11	0.834308	100/100	BlockFrequency
4	12	14	13	10	8	7	7	17	8	0.122325	100/100	CumulativeSums
10	9	10	15	9	12	9	8	9	9	0.924076	100/100	Runs
9	14	7	8	11	12	15	10	7	7	0.554420	100/100	LongestRun
10	11	11	4	14	13	8	13	11	5	0.334538	100/100	Rank
14	9	13	7	11	7	11	14	8	6	0.514124	99/100	FFT
6	10	10	11	5	18	12	3	9	16	0.020548	100/100	NonOverlappingTemplate
12	13	14	11	8	7	9	10	6	10	0.739918	100/100	OverlappingTemplate
7	12	13	16	11	13	13	5	5	5	0.085587	99/100	Universal
12	12	15	4	11	7	10	8	6	15	0.191687	100/100	ApproximateEntropy
3	9	7	7	10	7	6	3	6	6	0.568055	64/64	RandomExcursions
1	6	5	8	8	5	6	7	8	10	0.407091	64/64	RandomExcursionsVariant
14	8	11	10	11	14	9	2	9	12	0.289667	100/100	Serial
9	10	10	5	16	8	5	12	13	12	0.289667	100/100	LinearComplexity

a)

Figure 12. $NTTL_2$ map successfully passed NIST tests

We have explored several topologies and finally proposed two new 2-D CPRNG. The proposed models with injection mechanism allow to puzzle perfectly the pieces of the chaotic attractor, like a true random generator. To achieve the best distribution in the phase space, the modified form $MTTL_2^{SC}$ alternate map has been proposed. The new map exhibits excellent features due to the injection mechanism and enables the uniform density in the state space. The system exhibits strong nonlinear dynamics, demonstrating strong sensitivity to initial conditions. It generates an infinite range of intensive chaotic behavior with large positive Lyapunov exponent values. Moreover, $MTTL_2^{SC}$ successfully passed all required tests: cross-correlation, autocorrelation, LLE, NIST tests, uniform attractor on the phase space and phase delay. The system analysis and the dynamics evolution by bifurcation diagram and topological mixing proved the complex behavior. The system orbits exhibited complex behavior with perfect mixing. The study demonstrated totally unpredictable (for any intruder) dynamics making the system strong-potential candidate for high-security applications. Another CPRNG candidate based on $NTTL_2^{SC}$ map was proposed that successfully passed all required statistical, graphical and numerical results for both states components. The $NTTL_2^{SC}$ map demonstrates complex dynamics being very promising to real scale cryptography application.

References

[1] Menezes, A. J., Van Oorschot, P. C., and Vanstone, S. A. (1996) CRC press *Handbook of applied cryptography*

[2] Li, C. Y., Chen, Y. H., Chang, T. Y., Deng, L. Y., and To, K. (2012). Period extension and randomness enhancement using high-throughput reseeding-mixing PRNG. *Very Large Scale Integration (VLSI) Systems, IEEE Transactions on*, **20(2)**, 385-389.

[3] Lozi, R. (2012). Emergence of randomness from chaos, *International Journal of Bifurcation and Chaos*, **22 (02)**, 12500211/15.

[4] Nejati, H., Beirami, A., and Massoud, Y. (2008, August). A realizable modified tent map for true random number generation. *In Circuits and Systems*, MWSCAS 2008. 51st Midwest Symposium on (pp. 621-624). IEEE.

[5] May, R. M. (2001). *Stability and complexity in model ecosystems.* Princeton University Press, **Vol. 6**.

[6] May, R. M. (1974). Biological populations with nonoverlapping generations: stable points, stable cycles, and chaos. *Science*, **186(4164)**, 645-647.

[7] Feigenbaum, M. J. (1979). The universal metric properties of nonlinear transformations. *Journal of Statistical Physics*, **21(6)**, 669-706.

[8] Sudret, B. (2008). Global sensitivity analysis using polynomial chaos expansions. *Reliability Engineering and System Safety*, **93(7)**, 964-979.

[9] Lozi, R. (2013) Can we trust in numerical computations of chaotic solutions of dynamical systems? *In Topology and Dynamics of Chaos, Ch. Letellier, R. Gilmore (Eds.)*, World Scientific Series in Nonlinear Science Series A, Chapt. **3**, 2013.

[10] Ariffin, M. R. K., and Noorani, M. S. M. (2008). Modified Baptista type chaotic cryptosystem via matrix secret key. *Physics Letters A*, **372(33)**, 5427-5430.

[11] Lozi, R. (2013) Designing chaotic mathematical circuits for solving practical problems, *International Journal of Automation and Computing*, **vol. 11**, no. 6, pp. 588597, 2014.

[12] Yuan, G., and Yorke, J. A. (2000). Collapsing of chaos in one dimensional maps. *Physica D: Nonlinear Phenomena*, **136(1)**, 18-30.

[13] Lanford III, O. E. (1998). Informal remarks on the orbit structure of discrete approximations to chaotic maps. *Experimental Mathematics*, **7(4)**, 317-324.

[14] Rojas, A. E., Taralova, I., and Lozi, R. (2013). New alternate ring-coupled map for multi-random number generation, *Journal of Nonlinear Systems and Applications*, **4(1)**, 64-69.

[15] Garasym O. (2015). Application of nonlinear dynamics to the design of robust chaotic generators. *PhD thesis, Ecole Centrale de Nantes*, pp. 1-170.

[16] Lozi, R. (2009) Chaotic pseudo random number generators via ultra weak coupling of chaotic maps and double threshold sampling sequences *In proceedings of: ICCSA 2009 The 3rd International Conference on Complex Systems and Applications, University of Le Havre, France*, 20-24.

[17] Manjunath, G., Fournier-Prunaret, D., and Taha, A. K. (2008). A 3-dimensional piecewise affine map used as a chaotic generator *European Conference on Iteration Theory September(ECIT), Yalta, Ukraine*, pp. 7-13.

On the set of non-wandering points of monotone maps on local dendrites

E N Makhrova[1], **K S Vaniukova**[2]

Institute of Information Technology, Mathematics and Mechanics, Lobachevsky State University of Nizhni Novgorod, Nizhni Novgorod, Russia

E-mail: [1]`elena_makhrova@inbox.ru`, [2]`vaniukovaks@gmail.com`

Abstract. Let X be a local dendrite, $f : X \to X$ be a monotone map. In the article the structure of the set of non-wandering points of f is investigated.

1. Introduction

By *continuum* we mean a compact connected metric space. An *arc* is a set which is homeomorphic to the closed interval $[0, 1]$. A circle is denoted by S^1. A *closed curve* is any space homeomorphic to a circle.

Let X be a continuum and $A \subset X$. We denote by $\overline{A}$ the closure of A in X. The symbol $diam(A)$ means the diameter of A. The $Card(A)$ is the cardinality of A.

We use $\mathbf{N}$ to denote the set of positive integers.

A *dendrite* is a locally connected continuum without subsets homeomorphic to a circle. It is well known that for every two points in a dendrite there is a unique arc joining these points (see, e.g., [3, §51]). If X is a dendrite, then we denote by $[x, y]$ the arc joining and containing points x, y. Let $(x, y) = [x, y] \setminus \{x, y\}$, $(x, y] = [x, y] \setminus \{x\}$, $[x, y) = [x, y] \setminus \{y\}$.

A continuum X is called a *local dendrite* if for any point $x \in X$ there exists a neighborhood $U(x)$ of a point x such that $\overline{U(x)} \bigcap X$ is a dendrite. A local dendrite is a locally connected continuum containing a finite number of closed curves (see, e.g., [3, §51]). Hence any two different points in X can be joined by a finite number of arcs. Let x, y be any different points in X. We will denote by $\gamma_1[x, y], \gamma_2[x, y], ..., \gamma_n[x, y]$ the arcs joining and containing x, y. Let $\gamma_i(x, y) = \gamma_i[x, y] \backslash \{x, y\}$, $\gamma_i(x, y] = \gamma_i[x, y] \backslash \{x\}$ and $\gamma_i[x, y) = \gamma_i[x, y] \backslash \{y\}$. Every subcontinuum of a local dendrite is a local dendrite (see [3, §51]). A closed curve in a local dendrite is denoted by Θ.

Let X be a local dendrite with a metric d and a point $p \in X$. We use the notion of an order of a point in the sense of Menger-Urysohn (see, [3, §51]).

We say that an *order of a point* $p \in X$ *does not exceed* n ($ord\ p \leq n$) if for any $\varepsilon > 0$ there exists δ, $0 < \delta < \varepsilon$, such that $Card(\partial U_\delta(p)) \leq n$, where $\partial(\cdot)$ denotes the boundary of a set. The equality $ord\ p = n$ means that $ord\ p \leq n$ and the statement $ord\ p \leq m$ is false for any $m < n$. A point p is called to be:

- an *end point of* X if $ord\ p = 1$; denote by $E(X)$ the set of end points of X.
- a *branch point of* X if $ord\ p \geq 3$; denote by $R(X)$ the set of branch points of X.

Journal of Physics: Conference Series **692** (2016) 012012 doi:10.1088/1742-6596/692/1/012012

It is well known that $ord\ x \leq \omega$ for any $x \in X$ [3, §51]. The set $X \setminus \{p\}$ consists of one or more connected set. Each such set is called *a component at a point p*.

Let $f : X \to X$ be a continuous map of a local dendrite X. A map f is said to be *monotone* if for any connected subset C of X, $f^{-1}(C)$ is connected [3, §46]. A monotone map from a dendrite into itself is called *a monotone dendrite map*. A monotone map from a local dendrite into itself is called *a monotone local dendrite map*.

Let $f : X \to X$ be a monotone local dendrite map. Denote by f^n the n-iterate of f; that is, $f^0 =$ identity and $f^n = f \circ f^{n-1}$ if $n \geq 1$. Let a point $x \in X$. The set $\{x, f(x), f^2(x), ...\}$ is called *the orbit of a point x*. We define the ω-limit set of a point x to be the set

$$\omega(x, f) = \{z \in X : \exists n_j \in \mathbf{N},\ n_j \to \infty,\ \lim_{j\to\infty} d(f^{n_j}(x), z) = 0\}.$$

A point $x \in X$ is called:

- *a periodic point of period n*, $n \in \mathbf{N}$ if $f^n(x) = x$ and $f^i(x) \neq x$ for $1 \leq i \leq n-1$; denote by $Per(f)$ the set of periodic points of f. If $n = 1$ then x is called *a fixed point of f*; denote by $Fix(f)$ the set of fixed points of f. The orbit of a periodic point x under f is denoted by $Orb(x, f)$.
- *a non-wandering point of f* if for any neighborhood $U(x)$ of x there is $n \in \mathbf{N}$ such that $f^n(U(x)) \bigcap U(x) \neq \emptyset$. A set of non-wandering points of f is denoted by $\Omega(f)$.

It is well known that $Per(f) \subseteq \bigcup\limits_{x\in X} \omega(x, f) \subseteq \Omega(f)$.

A subset $A \subset X$ is called *f-invariant* if $f(A) \subseteq A$. It is called *minimal* for f if it is closed, non-empty, f-invariant and that does not contain any non-empty closed f-invariant proper subset.

In the article we will prove the next result.

Theorem 1.1 *Let $f : X \to X$ be a monotone local dendrite map. If $Per(f) = \emptyset$, then there is a closed curve $\Theta \subset X$ such that either $\Omega(f) = \Theta$ or $\Omega(f)$ is a Cantor set contained in Θ. If $Per(f) \neq \emptyset$, then $\Omega(f) = \overline{Per(f)}$.*

2. Auxiliary results

Lemma 2.1 [8],[1] *Let $f : S^1 \to S^1$ be a monotone map of a circle S^1. Then for any $x \in S^1$, $\omega(x, f)$ is a minimal set which is either finite, or a Cantor set, or a circle S^1.*

Lemma 2.2 [5] *Let $\{Y_n\}_{n\geq 1}$ be a sequence of connected subsets of a dendrite X. If $Y_i \cap Y_j = \emptyset$ for all $i \neq j$, then $\lim\limits_{n\to\infty} diam\,(Y_n) = 0$.*

Lemma 2.3 [2] *Let $f : X \to X$ be a continuous dendrite map. Then $Fix(f) \neq \emptyset$.*

Lemma 2.4 [6] *Let $f : X \to X$ be a monotone dendrite map. Then for any points $x, y \in X$, $f([x, y]) = [f(x), f(y)]$.*

Lemma 2.5 [7] *Let $f : X \to X$ be a monotone dendrite map, $a \in Fix(f)$ and $x \in X$ be such that $[a, x] \subset [a, f(x)]$. Then there exists a point $b \in Fix(f)$ such that $\lim\limits_{n\to\infty} f^n(x) = b$ and $[a, f^n(x)] \subset [a, b]$ for all $n \in \mathbf{N}$.*

Lemma 2.6 [7] *Let $f : X \to X$ be a monotone dendrite map. Then for any point $x \in X$, $\omega(x, f)$ is either a periodic orbit or a minimal Cantor set.*

Lemma 2.7 [4] *Let $f : X \to X$ be a monotone dendrite map and z be a branch point such that $z \notin Per(f)$. Then at most two components at a point z contain periodic points of f.*

Lemma 2.8 [4] *Let $f : X \to X$ be a monotone dendrite map with a sequence of periodic points $\{z_i\}_{i\geq 1}$ such that $\lim_{i\to\infty} z_i = a$ and all the points z_i, $i \geq 1$, lie in some component $X_1(a)$ at a point a. Then there exists a sequence $\{x_i\}_{i\geq 1} \subset Per(f) \cap X_1(a)$ such that $\{x_i\}_{i\geq 1} \subset [z_1, a]$ and $\lim_{i\to\infty} x_i = a$.*

Lemma 2.9 [4] *Let $f : X \to X$ be a monotone dendrite map. Then for any point $y \in X$ such that $ord\, y \geq 2$ and for different components $X_1(y)$, $X_2(y)$ at a point y, we have $f(\overline{X_1(y)}) \cap f(\overline{X_2(y)}) = \{f(y)\}$.*

Lemma 2.10 [1] *If $f : X \to X$ is an onto monotone local dendrite map and Θ is a closed curve in X, then $f^{-1}(\Theta)$ contains a closed curve in X.*

Lemma 2.11 [1] *Let X be a local dendrite and let A and B be two subcontinua of X such that $A \bigcap B$ is infinite. Then $A \bigcap B$ contains an arc of X.*

Lemma 2.12 [9, §16] *A non-empty set M is minimal if and only if $\omega(x, f) = M$ for every $x \in M$.*

Lemma 2.13 [9, §16] *Any closed invariant compact set has a minimal set.*

Lemma 2.14 [4] *Let $f : X \to X$ be a monotone dendrite map, a point $z \in Fix(f)$. If $x \in Per(f) \setminus Fix(f)$, then any arc $[z, y]$ contains at most one point of $Orb(x, f)$.*

Lemma 2.15 *Let $f : X \to X$ be a monotone local dendrite map. Then for any points $x, y \in X$ such that $f(x) \neq f(y)$ and any arc $\gamma_i[x, y]$ there is $\gamma_j[f(x), f(y)]$ such that $f(\gamma_i[x, y]) = \gamma_j[f(x), f(y)]$.*

Proof. Since f is continuous, then $f(\gamma_i[x, y])$ is connected. Hence, there is an arc $\gamma_j[f(x), f(y)]$ such that

$$\gamma_j[f(x), f(y)] \subseteq f(\gamma_i[x, y]). \tag{1}$$

On the other hand since f is monotone, then $f^{-1}(\gamma_j[f(x), f(y)])$ is connected and so $\gamma_i[x, y] \subseteq f^{-1}(\gamma_j[f(x), f(y)])$. Hence,

$$f(\gamma_i[x, y]) \subseteq f(f^{-1}(\gamma_j[f(x), f(y)])) = \gamma_j[f(x), f(y)]. \tag{2}$$

By (1) and (2), $f(\gamma_i[x, y]) = \gamma_j[f(x), f(y)]$. Lemma 2.15 is proved.

Lemma 2.16 *Let $f : X \to X$ be a monotone local dendrite map and $\Theta \subset X$ be a closed curve. Then $f(\Theta)$ is either a closed curve or a singleton.*

Proof. If $f(\Theta)$ is a singleton, then Lemma 2.16 is proved. Let $f(\Theta)$ is not a singleton. We choose different points $a, b \in \Theta$. Then there are two arcs $\gamma_1[a, b], \gamma_2[a, b] \subset \Theta$ such that

$$\gamma_1(a, b) \cap \gamma_2(a, b) = \emptyset. \tag{3}$$

By Lemma 2.15, $f(\gamma_i[a, b]) = \gamma_{j_i}[f(a), f(b)]$, for $1 \leq i \leq 2$. By (3) and monotonicity of f, $\gamma_{j_1}(f(a), f(b)) \cap \gamma_{j_2}(f(a), f(b)) = \emptyset$. It follows that $\gamma_{j_1}[f(a), f(b)] \cup \gamma_{j_2}[f(a), f(b)]$ is a closed curve. Lemma 2.16 is proved.

Lemma 2.17 *Let $f : S^1 \to S^1$ be a monotone map on a circle and $Per(f) = \emptyset$. Then $\Omega(f)$ is either S^1 or a minimal Cantor set.*

Journal of Physics: Conference Series **692** (2016) 012012 doi:10.1088/1742-6596/692/1/012012

Proof. Let x be any point from S^1 and let $M = \omega(x, f)$. Since $Per(f) = \emptyset$, then by Lemma 2.1, M is either S^1 or minimal Cantor set. It is evident that $M \subseteq \Omega(f)$. We show that $\Omega(f) = M$. The statement is true if $M = S^1$. Now let M is a Cantor set and let an arc $\gamma_1(\alpha, \beta) \subset S^1 \backslash M$ be such that $\alpha, \beta \in M$. We show that $f^i(\gamma_1(\alpha, \beta)) \cap \gamma_1(\alpha, \beta) = \emptyset$ for any $i \in \mathbf{N}$. Suppose on the contrary. Then there are a point $z \in \gamma_1(\alpha, \beta)$ and a number $I \in \mathbf{N}$ such that $f^I(z) \in \gamma_1(\alpha, \beta)$. Since $\alpha, \beta \in M$ and $f(M) = M$, then $f^I(\alpha), f^I(\beta) \notin (\alpha, \beta)$. It follows from here, Lemma 2.15 and the fact that $f^I(\gamma_1(\alpha, \beta)) \cap \gamma_1(\alpha, \beta) \neq \emptyset$, that $\gamma_1(\alpha, \beta) \subset f^I(\gamma_1(\alpha, \beta))$. Since f is monotone, then $\gamma_1(\alpha, \beta) \cap Fix(f^I) \neq \emptyset$, a contradiction. Thus, $f^i(\gamma_1(\alpha, \beta)) \cap \gamma_1(\alpha, \beta) = \emptyset$ for any $i \in \mathbf{N}$. It follows that $\Omega(f) \cap \gamma_1(\alpha, \beta) = \emptyset$. Since $\gamma_1(\alpha, \beta)$ is any arc in $S^1 \backslash M$, then $\Omega(f) = M$. Lemma 2.17 is proved.

The next statement is easy to prove.

Lemma 2.18 *Let $f : S^1 \to S^1$ be a monotone map on a circle and $Per(f) \neq \emptyset$. Then $\Omega(f) = Per(f)$.*

3. Proof of Theorem 1.1

We need some auxiliary results.

Lemma 3.1 *Let $f : X \to X$ be a monotone local dendrite map. Assume $\Theta \subset X$ is a closed curve such that $f(\Theta) = \Theta$ and $\Theta \bigcap Per(f) = \emptyset$. Then for any non-empty connected component $V \subset X \backslash \Theta$ there exists $I \in \mathbf{N}$ such that $f^i(V) \bigcap V = \emptyset$ for any $i \geq I$.*

Proof. Let $A = \overline{V} \bigcap \Theta$. We claim that A is finite. If not, then by Lemma 2.11, A contains an arc. On the other hand $A \subsetneq R(X)$ and so $R(X)$ is uncountable, a contradiction. Thus, A is finite. Let $I = Card(A)$. Since $A \subset \Theta$, then $Per(f) \bigcap A = \emptyset$. Then

$$f^i(a) \notin A, \text{ for all } i \geq I \text{ and for any } a \in A. \tag{4}$$

We show that $f^i(V) \bigcap V = \emptyset$ for any $i \geq I$. Suppose on the contrary. Then there are $x \in V$ and a natural number $J \geq I$ such that $f^J(x) \in V$. Let $a \in A$. Since $\overline{V}$ is a local dendrite, then there is an arc $\gamma_1[a, x] \subset \overline{V}$. Since $f(\Theta) = \Theta$ and $a \in \Theta$, then by (4), $f^J(a) \notin \overline{V}$. But $f^J(x) \in V$. Hence, $A \cap f^J(\gamma_1(a, x)) \neq \emptyset$ and so there is a point $b \in A$ such that $b \in f^J(\gamma_1(a, x))$. Therefore, there exists a point $z_1 \in \gamma_1(a, x) \subset V$ such that $f^J(z_1) = b$. On the other hand $f^J(\Theta) = \Theta$. Hence, there is $z_2 \in \Theta$ such that $f^J(z_2) = b$. Since $\Theta \cap Per(f) = \emptyset$, then $z_2 \neq b$ and $f^{-J}(b) \cap \{b\} = \emptyset$. Hence, $f^{-J}(b)$ is not connected, a contradiction. Thus, $f^i(V) \bigcap V = \emptyset$ for any $i \geq I$. Lemma 3.1 is proved.

Lemma 3.2 *Let $f : X \to X$ be a monotone local dendrite map. Assume $\Theta \subset X$ is a closed curve such that $f(\Theta) = \Theta$ and $\Theta \bigcap Per(f) = \emptyset$. Then either $\Omega(f) = \Theta$ or $\Omega(f)$ is a Cantor set contained in Θ.*

Proof. By Lemma 3.1,

$$\Omega(f) \subseteq \Theta. \tag{5}$$

According to Lemma 2.17, $\Omega(f|_\Theta)$ is either Θ or a Cantor set. If $\Omega(f|_\Theta) = \Theta$, then by (5), $\Omega(f) = \Theta$ and Lemma 3.2 is proved.

Let $\Omega(f|_\Theta)$ be a Cantor set. We show that any point $x \in \Theta \setminus \Omega(f|_\Theta)$ is wandering under f. Suppose on the contrary. Then there are sequences of points $\{a_i\}_{i \geq 1}$ and of numbers $\{n_i\}_{i \geq 1}$ such that $\lim_{i \to \infty} a_i = \lim_{i \to \infty} f^{n_i}(a_i) = x$. Since $x \notin \Omega(f|_\Theta)$, then there is $I \in \mathbf{N}$ so that $a_i \notin \Theta$ for any $i \geq I$. Hence for all $i \geq I$ there are a point $r_i \in \Theta \cap R(X)$ and an arc $\gamma_1[r_i, a_i]$ such that

Journal of Physics: Conference Series **692** (2016) 012012 doi:10.1088/1742-6596/692/1/012012

$\gamma_1(r_i, a_i) \cap \Theta = \emptyset$. Since X is a locally connected then $\lim_{i\to\infty} r_i = x$. Since $x \notin \Omega(f|_\Theta)$, there is an arc $\gamma_1(\alpha,\beta) \subset \Theta$ containing x such that

$$f^i(\gamma_1(\alpha,\beta)) \cap \gamma_1(\alpha,\beta) = \emptyset, \quad \text{for any } i \in \mathbf{N}. \tag{6}$$

We take a connected neighborhood $U(x)$ of a point x so that $U(x) \cap \Theta \subset \gamma_1(\alpha,\beta)$. There is a natural number $J \geq I$ such that $r_J, f^{n_J}(a_J) \in U(x)$. It follows from here and (6) that $f^J(\gamma_1[r_J, a_J]) \cap \gamma_1(\alpha,\beta)$ is an arc. Let an arc $\gamma_1[b_1,b_2] \subset f^J(\gamma_1(r_J,a_J)) \cap \gamma_1(\alpha,\beta)$. By continuity of f, there is an arc $\gamma_1[a_1,a_2] \subset \gamma_1(r_J,a_J)$ such that $f(\gamma_1[a_1,a_2]) = \gamma_1[b_1,b_2]$. On the other hand $f^J(\Theta) = \Theta$. Hence, there is an arc $\gamma_1[c_1,c_2] \subset \Theta$ so that $f^J(\gamma_1[c_1,c_2]) = \gamma_1[b_1,b_2]$. By (6), $\gamma_1[c_1,c_2] \cap \gamma_1(\alpha,\beta) = \emptyset$. Thus, $f^{-J}(\gamma_1[b_1,b_2]) \cap \gamma_1(\alpha,\beta) = \emptyset$ and so $f^{-J}(\gamma_1[b_1,b_2])$ is not connected, a contradiction. Thus, $\Omega(f) = \Omega(f|_\Theta)$. Lemma 3.2 is proved.

Repeating arguments of the proof of Lemma 3.2 and using Lemma 2.18, it is easy to prove the next statement.

Lemma 3.3 *Let $f : X \to X$ be a monotone local dendrite map. Assume $\Theta \subset X$ is a closed curve such that $f(\Theta) = \Theta$ and $\Theta \cap Per(f) \neq \emptyset$. Then $\Omega(f|_\Theta) = \Omega(f) \cap \Theta$.*

Lemma 3.4 *Let $f : X \to X$ be a monotone local dendrite map and $\Theta \subset X$ be a closed curve. If there is natural number $n \geq 2$ such that $f^n(\Theta) = \Theta$, then $\Theta \bigcap Per(f) \neq \emptyset$.*

Proof. Suppose that the statement is false. As $n \geq 2$, then $f(\Theta) \neq \Theta$. By Lemma 2.16, $f(\Theta)$ is either a closed curve or a singleton. Since $f^n(\Theta) = \Theta$, then $f(\Theta)$ is a closed curve. Moreover,

$$f^{ni}(f(\Theta)) = f(\Theta), \quad \text{for any } i \in \mathbf{N}. \tag{7}$$

Since $f(\Theta) \neq \Theta$, then there is a non-empty connected component $V \subset X \backslash \Theta$ such that $V \bigcap f(\Theta) \neq \emptyset$. By Lemma 3.1, there is $I \in \mathbf{N}$ such that $f^i(V) \bigcap V = \emptyset$ for any $i \geq I$. Hence, $f^i(f(\Theta)) \neq f(\Theta)$ for any $i \geq I$. This contradicts (7). Lemma 3.4 is proved.

Lemma 3.5 *Let $f : X \to X$ be a monotone local dendrite map. If for any closed curve $\Theta \subset X$ and for any $n \in \mathbf{N}$, $f^n(\Theta) \neq \Theta$, then there is $I \in \mathbf{N}$ such that $f^I(X)$ is a dendrite.*

Proof. Let X has k closed curves and let $B = \{\Theta_1, ..., \Theta_k\}$. Since B is finite and for any closed curve $\Theta_i \subset B$, $f^n(\Theta_i) \neq \Theta_i$ for all $n \in \mathbf{N}$, then B is not invariant. Then there is Θ_j, $1 \leq j \leq k$, such that $f(\Theta_j) \notin B$. Hence, by Lemma 2.16, $f(\Theta_j)$ is a singleton. By Lemma 2.10, $f(X)$ contains at most $(k-1)$ closed curves. If $f(X)$ is a dendrite, then Lemma 3.5 is proved. Otherwise, $f(X)$ satisfies the conditions of Lemma 3.5. Repeating the above mentioned arguments we conclude that there is a natural number I such that $f^I(X)$ does not contain any closed curve. Thus, $f^I(X)$ is a dendrite. Lemma 3.5 is proved.

Let X be a dendrite. Take different points $a, b \in X$. Denote by $X_a(b)$ $(X_b(a))$ the component of a point b (a) containing a point a (b). Let $X_{(a,b)} = X_a(b) \cap X_b(a)$. Note that $(a,b) \subseteq X_{(a,b)}$. By 3-od we mean the set $\{z \in \mathbf{C} : z^3 \in [0,1]\}$, where $\mathbf{C}$ is the complex plane.

Lemma 3.6 *Let $f : X \to X$ be a monotone dendrite map, $a \in Fix(f)$ and a component $X(a)$ at a point a be such that $X(a) \cap Per(f) = \emptyset$. Then $X(a) \cap \Omega(f) = \emptyset$.*

Proof. The statement of Lemma 3.6 is true, if $f^n(X(a)) \cap X(a) = \emptyset$ for any $n \in \mathbf{N}$. Let $f^N(X(a)) \cap X(a) \neq \emptyset$ for some natural number $N \geq 1$. Put $g = f^N$.

Journal of Physics: Conference Series **692** (2016) 012012 doi:10.1088/1742-6596/692/1/012012

We claim that for any point $x \in X(a)$ there is $I \in \mathbf{N}$ such that $g^{-I}(x) = \emptyset$. Suppose on the contrary. Then $g^{-i}(x) \neq \emptyset$ for some point $x \in X(a)$ and for any $i \geq 1$. Since $g(X(a)) \cap X(a) \neq \emptyset$, then by Lemma 2.9, $g^{-1}(X(a)) \subseteq X(a)$ and so

$$g^{-i}(x) \subset X(a), \text{ for any } i \in \mathbf{N}. \tag{8}$$

According to Lemma 2.5,

$$g^{-i}(x) \cap \gamma(a, x) = \emptyset, \text{ for any } i \in \mathbf{N}. \tag{9}$$

We take any point $x_i \in g^{-i}(x)$ for every $i \in \mathbf{N}$. Denote by A the set of limit points of $\{x_i\}_{i \geq 1}$. Then A is closed and so A is compact. It is easy to show that A is an invariant set. By Lemma 2.13, there is a minimal set $M \subseteq A$. According to Lemma 2.12, $M = \omega(x, g)$ for any $x \in M$. By Lemma 2.6, M is either a periodic orbit or a Cantor set. We consider the following cases:

Case 1. M is a periodic orbit.

By (8), $M \subset \overline{X(a)}$. Since $X(a) \cap Per(f) = \emptyset$, then $M = \{a\}$ and so $\lim_{i \to \infty} x_i = a$. Hence there is $I \in \mathbf{N}$ such that $x_i \in X_{(a,x)}$ for any $i \geq I$. Using (9), we have that for any $i \geq I$ there are a branch point $r_i \in (a, x)$ and its component $X(r_i)$ such that $X(r_i) \cap (a, x) = \emptyset$ and $x_i \in X(r_i)$. Since $\lim_{i \to \infty} x_i = a$, then $\lim_{i \to \infty} r_i = a$. As $g^{-i}(x)$ is connected, then by (9), $g^{-i}(x) \subset X(r_i)$ for $i \geq I$. Hence $g(x_{i+1}) \subset X(r_i)$ and so

$$g^i(x_{2i}) \in X(r_i), \text{ for any } i \geq I. \tag{10}$$

By Lemma 2.5 and the fact that $X(a) \cap Per(g) = \emptyset$, we have

$$g^i(r_i) \in [a, r_i) \text{ for any } i \geq I. \tag{11}$$

We take a natural number $J \geq I$ such that $r_{2J} \in (a, r_J)$. It follows from (10) and (11), that $r_J \in g^J([r_{2J}, x_{2J}]) \cap g^J([r_J, x_J])$. Since $[r_{2J}, x_{2J}] \cap [r_J, x_J] = \emptyset$, then $g^{-J}(r_J)$ is not connected, a contradiction.

Case 2. M is a Cantor set. Then $M \cap \{a\} = \emptyset$ and so

$$M \subset X(a). \tag{12}$$

Let x, y be any different points in M. We show that

$$M \cap (a, y) = \emptyset \text{ for any point } y \in M. \tag{13}$$

Suppose on the contrary, that is, there is a point $x \in M \cap (a, y)$. Let $X_y(x)$ be the component of x containing y. Note that $X_y(x) \subset X(a)$. Since $y \in \omega(x, g)$, then there exists $n \in \mathbf{N}$ with $g^n(x) \in X_y(x)$. Then $x \in (a, g^n(x))$ and by Lemma 2.5, $Fix(g^n) \cap X(a) \neq \emptyset$, a contradiction. Thus, (13) is true.

By (13), $[a, x] \cup [a, y]$ is homeomorphic to 3-od for any points $x, y \in M$. Let r be a branch point of $[a, x] \cup [a, y]$. Denote by $X_y(r)$, $X_x(r)$, $X_a(r)$ the components at a point r containing y, x, a, respectively. Again there is $n \in \mathbf{N}$ such that $g^n(x) \in X_y(r)$. Then points $r, g^n(r) \in [a, g^n(x)]$. By Lemma 2.5 and the fact that $X(a) \cap Per(f) = \emptyset$, we have $g^n(r) \in [a, r)$. Thus, $g^n|_{[a,r]}$ is a monotone map with only one fixed point a. Hence $g^{ni}(r) \in [a, r)$ for any $i \in \mathbf{N}$. We show that $g^{ni}(r) \neq a$. Suppose that $g^{nI}(r) = a$ for some $I \in \mathbf{N}$. Since $X_x(r) \cap M \neq \emptyset$ and $X_y(r) \cap M \neq \emptyset$, then by (12), $g^{ni}(X_x(r)) \cap X(a) \neq \emptyset$ and $g^{ni}(X_y(r)) \cap X(a) \neq \emptyset$ for any $i \in \mathbf{N}$. Then $g^{nI}(X_x(r)) \cap g^{nI}(X_y(r)) \neq \{g^{nI}(r)\} = \{a\}$. This contradicts Lemma 2.9. Thus, $g^{ni}(r) \in (a, r)$ for any $i \in \mathbf{N}$. Moreover, $g^{n(i+1)}(r) \in (a, g^{ni}(r))$ and $\lim_{i \to \infty} g^{ni}(r) = a$ for any $i \in \mathbf{N}$.

By Lemma 2.9,
$$g^{ni}(\overline{X_a(r)}) \cap g^{ni}(\overline{X_y(r)}) \cap g^{ni}(\overline{X_x(r)}) = \{g^{ni}(r)\}.$$
Since $a \in Fix(f)$, then $g^{ni}(X_a(r)) \cap X_a(r) \neq \emptyset$. Since $g^{ni}(X_y(r)) \cap M \neq \emptyset$ and $g^{ni}(X_x(r)) \cap M \neq \emptyset$, then by (13), $g^{ni}(X_y(r)) \neq \{g^{ni}(r)\}$ and $g^{ni}(X_x(r)) \neq \{g^{ni}(r)\}$, for any $i \in \mathbf{N}$. Hence there is a component $X(g^{ni}(r))$ at a point $g^{ni}(r)$ so that $X(g^{ni}(r)) \cap (a, x) = \emptyset$ and $X(g^{ni}(r)) \cap M \neq \emptyset$. Note that $X(g^{ni}(r)) \cap X(g^{nj}(r)) = \emptyset$ for any different $i, j \in \mathbf{N}$. By Lemma 2.2, $\lim\limits_{i\to\infty} diam\,(X(g^{ni}(r)) = 0$. It means that $a \in M$, a contradiction.

Thus, for any point $x \in X(a)$ there is $I \in \mathbf{N}$ such that $g^{-I}(x) = \emptyset$. Hence $f^{-IN}(x) = \emptyset$. Since $f^{IN}(X)$ is compact, there is a neighborhood $U(x)$ of a point x so that $U(x) \subset X \setminus f^{IN}(X)$. It follows that $x \notin \Omega(f)$. Hence $\Omega(f) \cap X(a) = \emptyset$. Lemma 3.6 is proved.

Lemma 3.7 *Let $f : X \to X$ be a monotone dendrite map with different fixed points $a, b \in X$ such that $X_{(a,b)} \cap Per(f) = \emptyset$. Then $X_{(a,b)} \cap \Omega(f) = \emptyset$.*

Proof. Since $f|_{[a,b]}$ is a monotone map with only two fixed points a and b then one of them is an attracting point. Let a be such point. Then for any point $x \in (a, b)$, $f(x) \in [a, x)$. Hence for any point $x \in (a, b)$ there is an arc $(c, d) \subset (a, b)$ containing x such that
$$f^i([c, d]) \cap [c, d] = \emptyset \text{ for any } i \in \mathbf{N}. \tag{14}$$
Thus, if $X_{(a,b)} = (a, b)$, then Lemma 3.7 is proved. Now suppose that $X_{(a,b)} \neq (a, b)$. Then $(a, b) \cap R(X) \neq \emptyset$. Let r be any branch point in (a, b) and let $X(r)$ be any component at a point r such that $X(r) \cap (a, b) = \emptyset$. We take an arc $(c, d) \subset (a, b)$ such that $r \in (c, d)$ and (c, d) satisfies (14). We show that
$$f^i(X(r)) \cap X_{(c,d)} = \emptyset \text{ for any } i \in \mathbf{N}. \tag{15}$$
Suppose on the contrary. Then there are a point $y \in X(r)$ and a natural number I such that $f^I(y) \in X_{(c,d)}$. Without loss of generality we may assume that $c \in (a, r)$. By (14), $f^I(r) \in [a, c)$. Then $c \in f^I([r, y]) = [f^I(r), f^I(y)]$. Hence there exists a point $z_1 \in X(r)$ such that $f^I(z_1) = c$. On the other hand $f^I([a, b]) = [a, b]$. So there is a point $z_2 \in (a, b)$ such that $f^I(z_2) = c$. By (14), $z_2 \neq r$. It follows from here that $f^{-I}(c)$ is not connected, a contradiction. Thus, (15) is true.

By (14) and (15), $X_{(a,b)} \cap \Omega(f) = \emptyset$. Lemma 3.7 is proved.

Lemma 3.8 *Let $f : X \to X$ be a monotone dendrite map. Then $\Omega(f) = \overline{Per(f)}$.*

Proof. It is evident that $\overline{Per(f)} \subseteq \Omega(f)$. We prove the reverse inclusion. If $X = \overline{Per(f)}$, then the statement of Lemma 3.8 is true. Let $X \neq \overline{Per(f)}$. Then there is a nonempty component $V \subset X \setminus \overline{Per(f)}$. We show that $V \cap \Omega(f) = \emptyset$. By Lemma 2.7, $1 \leq Card\,(\overline{V} \cap \overline{Per(f)}) \leq 2$. We consider two cases.

Case 1. $Card\,(\overline{V} \cap \overline{Per(f)}) = 1$. Let $\overline{V} \cap \overline{Per(f)} = \{a\}$. We distinguish the following subcases.

Case 1.1. $a \in Per(f)$. Let m be period of a. By Lemma 3.6, $\Omega(f^m) \cap V = \emptyset$. If $m = 1$, Lemma 3.8 is proved. Let $m > 1$. We show that $f^i(V) \cap V = \emptyset$ for any $1 \leq i \leq m - 1$. If not then there are a point $x \in V$ and a number $1 \leq I \leq m - 1$ such that $f^I(x) \in V$. Since $V \cap Per(f) = \emptyset$ and $I \leq m - 1$, then $f^I(a) \notin \overline{V}$. Hence $a \in f^I((a, x))$ and there is a point $z_1 \in V$ so that $f^I(z_1) = a$. On the other hand there is a point $z_2 \in Orb(a, f)$ so that $f^I(z_2) = a$. Hence $f^{-I}(a)$ is not connected, a contradiction. Thus, $f^i(V) \cap V = \emptyset$ for any $1 \leq i \leq m-1$ and so $V \cap \Omega(f) = \emptyset$.

Case 1.2. $a \notin Per(f)$. We show that
$$f^i(a) \notin V \text{ for any } i \in \mathbf{N}. \tag{16}$$

Suppose that $f^I(a) \in V$ for some $I \in \mathbf{N}$. Since $a \in \overline{Per(f)}$, then by continuity of f, $f^I(Per(f)) \cap V \neq \emptyset$, a contradiction.

We show that $f^i(V) \cap V = \emptyset$ for any $i \in \mathbf{N}$. Suppose that there are a point $x \in V$ and a number $I \in \mathbf{N}$ such that $f^I(x) \in V$. By (16) and the fact that $a \notin Per(f)$, $f^I(a) \notin \overline{V}$. Hence, $a \in f^I((a,x))$. So there is a point $z_1 \in (a,x)$ such that $f^I(z_1) = a$. On the other hand since $a \in \overline{Per(f)} \setminus Per(f)$, then there is a point $z_2 \in \overline{Per(f)} \setminus Per(f)$ such that $f^I(z_2) = a$. Since $\overline{V} \cap \overline{Per(f)} = \{a\}$, then $z_2 \notin \overline{V}$. Since $a \notin Per(f)$, then $a \notin f^{-I}(a)$ and so $f^{-I}(a)$ is not connected. This contradicts monotonicity of f. Thus, $f^i(V) \cap V = \emptyset$ for any $i \in \mathbf{N}$. It follows from here that $V \cap \Omega(f) = \emptyset$.

Case 2. $Card\,(\overline{V} \cap \overline{Per(f)}) = 2$. Let points $a, b \in \overline{V} \cap \overline{Per(f)}$.

We show that $a, b \in Per(f)$. Since $V \cap Per(f) = \emptyset$, there are periodic points $x, y \in X$ such that $a, b \in [x, y]$. Let m be the least common multiple of periods of points x and y. Then a restriction $f^m|_{[x,y]}$ be a monotone map with $f(x) = x$, $f(y) = y$. Hence

$$Per(f^m|_{[x,y]}) = Fix(f^m|_{[x,y]}). \tag{17}$$

Suppose that $a \notin Per(f)$. Then $a \in (x, y)$. Hence there are two different components $X_x(a)$ and $X_y(a)$ at a point a containing periodic points x and y, respectively. It follows from Lemma 2.7, $Per(f) \subset X_x(a) \cup X_y(a)$. Since $a \in \overline{Per(f)}$, then there is a sequence of periodic points $\{z_i\}_{i\geq 1}$ such that $\lim\limits_{i\to\infty} z_i = a$. Then $\{z_i\}_{i\geq 1} \subset X_x(a) \cup X_y(a)$. Without loss of generality we may assume that $\{z_i\}_{i\geq 1} \subset X_x(a)$. By Lemma 2.8, there is a sequence $\{r_i\}_{i\geq 1} \subset Per(f) \cup X_x(a)$ such that $r_i \in [z_1, a]$ and $\lim\limits_{i\to\infty} r_i = a$ for any $i \in \mathbf{N}$. Since points $x, z_1 \in X_x(a)$, there is a natural number I such that $r_i \in [x, a]$ for any $i \geq I$. According to (17), $r_i \in Fix(f^m)$ for any $i \geq I$. By continuity of f, $f^m(a) = a$. This contradicts the assumption. Thus, $a \in Per(f)$. Analogously, $b \in Per(f)$.

Let k be the least common multiple of periods of points a and b. According to Lemma 3.7, $V \cap \Omega(f^k) = \emptyset$. Thus, if $k = 1$, then Lemma 3.8 is proved. Let $k > 1$. Since $V \cap Per(f) = \emptyset$, then points $f^i(a), f^i(b) \notin V$ and $a, b \notin [f^i(a), f^i(b)]$ for any $1 \leq i \leq k-1$. Hence

$$f^i([a,b]) \cap V = \emptyset, \text{ for any } 1 \leq i \leq k-1. \tag{18}$$

We show that $f^i(V) \cap V = \emptyset$ for any $1 \leq i \leq k-1$. Suppose on the contrary. Then for some $1 \leq I \leq k-1$, $f^I(V) \cap V \neq \emptyset$. We consider two cases.

Case 2.1. $f^I(a) = a$ or $f^I(b) = b$. Let $f^I(a) = a$. Denote by $X(a)$ the component at a point a containing V. Since $f^I(V) \cap V \neq \emptyset$, then $f^I(X(a)) \cap X(a) \neq \emptyset$. It follows from Lemma 2.9, $f^{-I}(X(a)) \subseteq X(a)$. Hence $Orb(b, f^I) \subset X(a)$. Note that $b \notin Fix(f^I)$, since $I < k$. Hence there is a point $c \in Orb(b, f^I) \cap X(a)$ such that $f^I(c) = b$. Since $V \cap Per(f) = \emptyset$, then $c \in X(a) \setminus V$. It follows that $b \in (a, c)$. This contradicts Lemma 2.14. Similarly, if $f^I(b) = b$.

Case 2.2. $f^I(a) \neq a$ and $f^I(b) \neq b$. Let a point $y \in V$ be such that $f^I(y) \in V$. By (18), $y \notin (a, b)$. Then there are a point $r \in (a, b)$ and a component $X(r)$ at a point r such that $X(r) \cap (a, b) = \emptyset$ and $y \in X(r)$. By Lemma 2.4, $f^I(r) \in f^I([a, b]) = [f^I(a), f^I(b)]$. By (18) and the fact that $f^I(y) \in V$, we have $(f^I(r), f^I(y)) \cap \{a, b\} \neq \emptyset$. Without loss of generality we may assume that $a \in (f^I(r), f^I(y))$. Then there is a point $z_1 \in (r, y)$ so that $f^I(z_1) = a$. On the other hand there is a point $z_2 \in Orb(a, f)$ such that $f^I(z_2) = a$. Since $a \notin f^{-I}(a)$, then $f^{-I}(a)$ is not connected, a contradiction.

Thus, $f^i(V) \cap V = \emptyset$ for any $1 \leq i \leq k-1$. It follows that $V \cap \Omega(f) = \emptyset$. Lemma 3.8 is proved.

Lemma 3.9 *Let X be a local dendrite with only one closed curve Θ, $f : X \to X$ be a monotone map such that $f(\Theta) = \Theta$ and $\Theta \cap Per(f) \neq \emptyset$. Then $\Omega(f) = \overline{Per(f)}$.*

Journal of Physics: Conference Series **692** (2016) 012012 doi:10.1088/1742-6596/692/1/012012

Proof. We take any point $a \in \Theta \cap R(X)$. Denote by V_Θ the connected component of the set $X \setminus \{a\}$ such that $\Theta \cap V_\Theta \neq \emptyset$. Then the set $Y = X \setminus V_\Theta$ is a dendrite. If $a \notin Per(f)$, then by monotonicity of f, $f^i(Y) \cap Y = \emptyset$ for any $i \in \mathbf{N}$. Hence $\Omega(f) \cap Y = \emptyset$.

Now let $a \in Per(f)$ of period $m \in \mathbf{N}$. Since $\Theta \subseteq f^m(\overline{V_\Theta})$ and $f^m(\Theta) = \Theta$, then by monotonicity of f, $f^m(Y) \cap V_\Theta = \emptyset$. Hence $f^m(Y) \subseteq Y$. According to Lemma 3.8, $\Omega(f^m|_Y) = \overline{Per(f^m|_Y)}$. Since a is a periodic point of period m and $f^i(\Theta) = \Theta$ for any $i \in \mathbf{N}$, then by monotonicity of f, $f^i(Y) \cap Y = \emptyset$ for $1 \leq i \leq m-1$. Hence $\Omega(f) \cap Y = \overline{Per(f)} \cap Y$. It follows from here, Lemma 2.18, Lemma 3.3 and the fact that a is any point in $\Theta \cap R(X)$, $\Omega(f) = \overline{Per(f)}$. Lemma 3.9 is proved.

Let X be a local dendrite, $\Theta \subset X$ be a closed curve and the set $Y = X \setminus \Theta$. Since $R(X)$ is at most countable, then $\overline{Y}$ consists of at most countable number of pairwise disjoint subcontinua.

Lemma 3.10 *Let $f : X \to X$ be a monotone local dendrite map, $\Theta \subset X$ be a closed curve, $Y = X \setminus \Theta$ and let Z be a subcontinuum in $\overline{Y}$.*
(3.10.1) Then the set $A = Z \cap \Theta$ is finite.
(3.10.2) If $f(Z) \cap Z \neq \emptyset$, then $f(Z) \subseteq Z$.

Proof. (3.10.1) If not, then by Lemma 2.11, A contains an arc and so $R(X)$ is uncountable, a contradiction.

(3.10.2) Suppose on the contrary. Then there is a point $x \in Z$ with $f(x) \notin Z$. Since $f(Z) \cap Z \neq \emptyset$, there is a point $y \in Z$ with $f(y) \in Z$. As Z is a local dendrite, there is an arc $\gamma_1[x, y] \subseteq Z$. Since $f(x) \notin Z$, then $f(\gamma_1[x, y]) \cap \Theta$ is a non-degenerate arc. As A is finite, then there is an arc $\gamma_1[b_1, b_2] \subset f(\gamma_1[x, y]) \cap \Theta$ such that

$$\gamma_1[b_1, b_2] \cap f(A) = \emptyset. \tag{19}$$

It follows that there exists an arc $\gamma_1[a_1, a_2] \subset \gamma_1[x, y] \setminus A$ and so $\gamma_1[a_1, a_2] \cap \Theta = \emptyset$. On the other hand $f(\Theta) = \Theta$. So there is an arc $\gamma_1[c_1, c_2] \subset \Theta$ such that $f(\gamma_1[c_1, c_2]) = \gamma_1[b_1, b_2]$. By (19), $f^{-1}([b_1, b_2]) \cap A = \emptyset$. So $f^{-1}([b_1, b_2])$ is not connected, a contradiction. Lemma 3.10 is proved.

Proof Theorem 1.1. 1. Let $Per(f) = \emptyset$. Then by Lemmas 3.5 and 2.3 there are a closed curve $\Theta \subset X$ and a natural number n such that $f^n(\Theta) = \Theta$. Since $\Theta \bigcap Per(f) = \emptyset$, then by Lemma 3.4, $n = 1$. By Lemma 3.2, either $\Omega(f) = \Theta$ or $\Omega(f)$ be a Cantor set contained in Θ. In this case, Theorem 1.1 is proved.

2. Let $Per(f) \neq \emptyset$. We consider two cases:

Case 1. There are a closed curve $\theta \subset X$ and $n \in \mathbf{N}$ such that $f^n(\Theta) = \Theta$.

Put $g = f^n$. We prove that $\Omega(g) = \overline{Per(g)}$. Let X has k closed curves. We proceed by induction. For $k = 1$ the statement is true by Lemma 3.9. We suppose that the statement is true for $k = n-1$ and we prove it for $k = n$. Let $Y = X \setminus \Theta$. Then $\overline{Y}$ can be written as a disjoint union of at most countable family of subcontinua. Let Z be one of them and let $A = Z \cap \Theta$. If $g^i(Z) \cap Z = \emptyset$ for all $i \in \mathbf{N}$, then $\Omega(g) \cap Z = \emptyset$. Let $g^I(Z) \cap Z \neq \emptyset$ for some $I \in \mathbf{N}$. If $I \geq 2$, then we suppose that

$$g^i(Z) \cap Z = \emptyset, \text{ for } 1 \leq i \leq I-1. \tag{20}$$

Since $g^I(Z) \cap Z \neq \emptyset$, then by Lemma 3.10, $g^I(Z) \subseteq Z$. It follows from here and the equality $g^I(\Theta) = \Theta$, that $g^I(A) \subseteq A$. By Lemma 3.10, A is finite. Hence $A \cap Per(f) \neq \emptyset$. Thus, Z is a continuum with at most $(n-1)$ closed curves and $Z \cap Per(f) \neq \emptyset$. By our assumption, $\Omega(g^I|_Z) = \overline{Per(g^I|_Z)}$. If $I \geq 2$, then by (20), $\Omega(g|_Z) = \overline{Per(g|_Z)}$.

It follows from here, Lemma 2.18 and Lemma 3.3, $\Omega(g) = \overline{Per(g)}$. Using this fact it is easy to prove that $\Omega(f) = \overline{Per(f)}$. In this case Theorem 1.1 is proved.

Case 2. For any closed curve $\Theta \subset X$ and for any $n \in \mathbf{N}$, $f^n(\Theta) \neq \Theta$. By Lemma 3.5, $f^I(X)$ is a dendrite for some $I \in \mathbf{N}$ and, by Lemma 3.8, we have $\Omega(f) = \overline{Per(f)}$. Theorem 1.1 is proved.

Journal of Physics: Conference Series **692** (2016) 012012 doi:10.1088/1742-6596/692/1/012012

Funding
This work was supported by the Ministry of Education and Science of the Russian Federation under grant N 1410.

References

[1] Abdelli H 2015 ω-limit sets for monotone local dendrite maps *Chaos, Solitons & Fractals* **71** 66-72
[2] Ayres W L 1930 Some generalizations of the Scherrer fixed-point theorem *Fund. Math.* **16** 332-6
[3] Kuratovski K 1968 *Topology, vol.2* (New York: Academic Press)
[4] Efremova L S and Makhrova E N 2001 The dynamics of monotone maps of dendrites *Sb. Math.* **192** 807-21
[5] Mai J and Shi E 2009 $\overline{R} = \overline{P}$ for maps of dendrites X with $Card(End(X)) < c$ *Int. J. Bifurcation and Chaos* **19** 1391–96
[6] Naghmouchi I 2011 Dynamics of monotone graph, dendrite and dendroid maps *I.J. Bifurcation & Chaos* **21** 3205–15
[7] Naghmouchi I 2012 Dynamical properties of monotone dendrite maps *Topology and its Applications* **159** 144–9
[8] Poincare H 1885 Sur les courbes définies par les équations différentielles *Journal de mathématiques pures et appliquées* **4** 167–244
[9] Sibirskii S K 1975 *Introduction to topological dynamics* (Netherlands: Springer)

Journal of Physics: Conference Series **692** (2016) 012013 doi:10.1088/1742-6596/692/1/012013

Using MatContM in the study of a nonlinear map in economics

Niels Neirynck [1], **Bashir Al-Hdaibat** [1], **Willy Govaerts** [1], **Yuri A. Kuznetsov** [2,3] and **Hil G E Meijer** [3]

[1] Ghent University, Ghent, Belgium
[2] Utrecht University, Utrecht, Netherlands
[3] University of Twente, Twente, Netherlands

E-mail: `Niels.Neirynck@UGent.be`

Abstract. MatContM is a MATLAB interactive toolbox for the numerical study of iterated smooth maps, their Lyapunov exponents, fixed points, and periodic, homoclinic and heteroclinic orbits as well as their stable and unstable invariant manifolds. The bifurcation analysis is based on continuation methods, tracing out solution manifolds of various types of objects while some of the parameters of the map vary. In particular, MatContM computes codimension 1 bifurcation curves of cycles and supports the computation of the normal form coefficients of their codimension two bifurcations, and allows branch switching from codimension 2 points to secondary curves. MatContM builds on an earlier command-line MATLAB package Cl_MatContM but provides new computational routines and functionalities, as well as a graphical user interface, enabling interactive control of all computations, data handling and archiving. We apply MatContM in our study of the monopoly model of T. Puu with cubic price and quadratic marginal cost functions. Using MatContM, we analyze the fixed points and their stability and we compute branches of solutions of period 5, 10, 13 17. The chaotic and periodic behavior of the monopoly model is further analyzed by computing the largest Lyapunov exponents.

1. Introduction

We consider dynamical systems generated by smooth nonlinear maps

$$x \mapsto f(x, \alpha), \quad x \in \mathbb{R}^n, \ \alpha \in \mathbb{R}^m \tag{1}$$

with state variable x and parameter vector α. The understanding of the dynamics of the map (1) requires most often numerical bifurcation analysis with the use of a dedicated software package. For this purpose we developed MatContM, a MATLAB continuation toolbox available at http://sourceforge.net/projects/matcont/. MatContM builds on the command line code Cl_MatContM described in [4], [5] and [6] but supports several new functionalities, as well as providing a uniform interface. A comprehensive tutorial is provided that illustrates the use of MatContM by investigating an example model. The tutorial can be found on http://.../cl_matcontm3p2/Tutorial_MatcontMGUI.pdf.

Experienced users can also use the code in MatContM via the MATLAB command line, which in some cases can extend the capabilities of the software.

MatContM computes bifurcations of fixed points and cycles in one parameter, continues bifurcation curves in two parameters and detects codimension 2 bifurcation points using

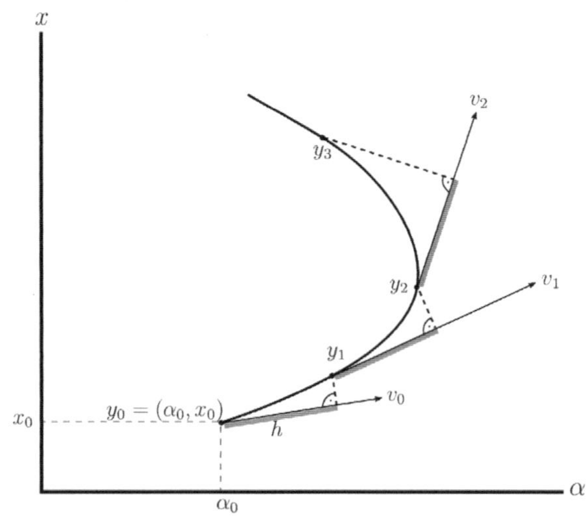

Figure 1: Schematic representation of pseudo-arclength continuation. The x represents the state space, the α represents the varying parameter, also referred to as the free parameter. MatContM uses a variant of Moore-Penrose continuation which builds upon pseudo-arclength continuation.

numerical continuation. A variant of pseudo-arclength continuation is used to approximate these curves (Figure 1). Bifurcations are detected by encountering zeros of test functions during the continuation process. The video file *animation1.mpg* demonstrates the continuation process for a fixed point continuation and *animation2.mpg* demonstrates how test functions are used to detect bifurcations (see Appendix A).

Critical normal form coefficients are computed at bifurcation points, using symbolic derivatives or if these are not available, finite differences or automatic differentiation is used. Automatic differentiation is usually slow, but becomes more competitive for cycles with high iteration number.

We note that there is another MATLAB interactive toolbox called MatCont [3] that provides numerical continuation and bifurcation algorithms for the study of parameterized continuous-time dynamical systems (ODEs).

2. Features and functionalities of MatContM

In a typical use of MatContM, one starts with an initial fixed point or cycle, which may be obtained from analysis, simulations or previous continuations. One first computes curves of fixed points or cycles under variation of one parameter, and may detect bifurcation points on such curves. Starting from such bifurcation points, the continuer algorithm in MatContM can compute bifurcation curves. These curves are defined by a system of equations consisting of fixed point and bifurcation conditions. With one free parameter we can also compute curves of connecting orbits. By varying two system parameters we can compute bifurcation curves of limit points, period-doubling and Neimark-Sacker points as well as tangencies of homoclinic and heteroclinic orbits. The systems that define connecting orbits and their tangencies are fairly complicated as they involve the saddle equilibria at the endpoints of the orbits, their eigenspace structures and the whole connecting orbit. The following list contains functionalities that are provided by MatContM:

- Simulation (iteration) of maps, i.e. computation and visualization of orbits (trajectories).

Journal of Physics: Conference Series **692** (2016) 012013 doi:10.1088/1742-6596/692/1/012013

- Computation of the Lyapunov exponents of long trajectories.
- Continuation of fixed points of maps and iterates of maps with respect to a control parameter.
- Detection of fold (limit point), flip (period-doubling point), Neimark-Sacker and branch points on curves of fixed points.
- Computation of normal form coefficients for fold, flip and Neimark-Sacker bifurcations.
- Continuation of fold, flip and Neimark-Sacker bifurcations in two control parameters. See *animation3.mpg* in Appendix A.
- Detection of all codimension 2 fixed point bifurcations on curves of fold, flip and Neimark-Sacker bifurcations.
- Computation of normal form coefficients for all codimension 2 bifurcations of fixed points.
- Switching to the period doubled branch in a flip point.
- Branch switching at branch points of fixed points.
- Switching to branches of codimension 1 bifurcations rooted in codimension 2 points. See *animation3.mpg* in Appendix A.
- Automatic differentiation for normal form coefficients of codimension 1 and codimension 2 bifurcations.
- Computation of one-dimensional invariant manifolds (stable and unstable) and in the two-dimensional case computing their transversal intersections to obtain initial homoclinic and heteroclinic connections.
- Continuation of homoclinic and heteroclinic orbits with respect to a control parameter and the detection of tangencies on the curve of orbits. See *animation4.mpg* in Appendix A.
- Continuation of homoclinic and heteroclinic tangencies in two control parameters. See *animation5.mpg* in Appendix A.

3. Practical use of MatContM

A simple but important feature is that MatContM has the ability to simulate maps, which means that it can compute orbits (trajectories). These orbits can also be visualized and their Lyapunov exponents can be computed.

The advanced use of MatContM relies on the ability to continue curves under variation of parameter(s) and apply advanced bifurcation theory on given examples of maps. The aim of the package is to allow a user to perform a bifurcation analysis of a map, without deep knowledge of the workings of MatContM or even the continuation software.

The user is able to enter a system, an initial fixed point and start computing with most of the settings left on default. This should lower the entry barrier for researchers from different research fields who want to investigate their models but do not want to be confronted with specification and implementation details.

The continuation curves can be visualized using the plot capabilities of MatContM; this can be done during and after the continuation.

Tools are provided to help with managing systems, diagrams and curves when generating large amount of data.

Specific features have been implemented to make branch switching between continuation curves fast and easy. An initial point can be selected out of a list of special points or it can be selected by double-clicking on a graph of the computed curve. A list of available curves for computation is shown depending on the type of the initial point.

An interface is also provided that allows the exchange of information between MatContM and the MATLAB command line. This enables the user to combine MatContM with other MATLAB software or simply to use computed data on the MATLAB command line.

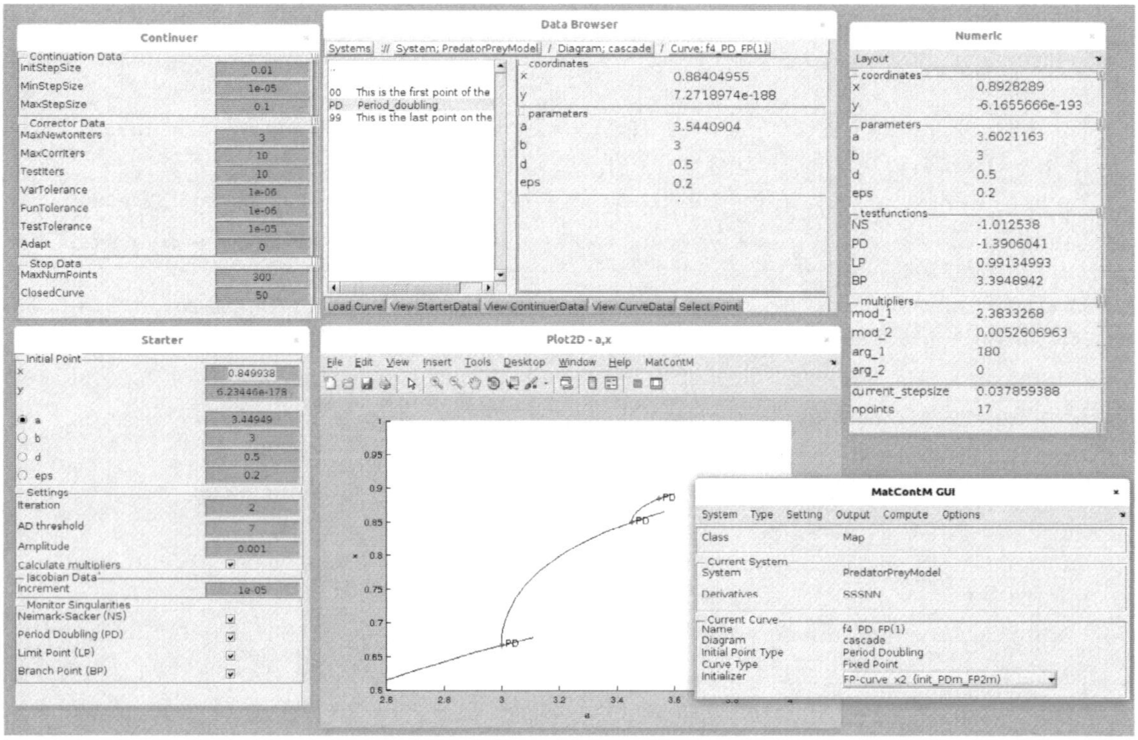

Figure 2: This action screenshot of MatContM presents the main windows that are opened during the computation of a cascade of period-doubling bifurcations in a predator-prey model [1]. Bottom right is the MatContM main window which contains general information on the model that is being used and the top-level commands. Top left is the Continuer window which shows the continuation variables. The fields of this window are independent of the particular curve that is being computed, though the numeric values can vary. Bottom left is the Starter window whose fields strongly depend on the curve that is being computed. Top middle is the Data Browser which allows to inspect all introduced systems, computed curves, etcetera. Top right is the Numeric window which during computation provides numerical values of computed quantities. Bottom middle is the 2D plot window which in this case plots several bifurcations curves; the horizontal axis presents a parameter of the system and the vertical axis is a state variable. The PD points are flip bifurcation points; the period-doubling cascade is clearly visible.

Figure 2 is an action screenshot of MatContM that gives a visual impression of the computation of a cascade of period-doubling bifurcations in a predator-prey model [1]. The video file *animation3.mpg* shows a two-parameter bifurcation analysis of the same model (Appendix A). The bifurcation diagram in *animation3.mpg* can be replicated using MatContM by following the tutorial available on the website.

4. Study of a nonlinear map in economics

We use MatContM to numerically analyze a monopoly model first suggested by Puu [7] which assumes a cubic price and quadratic marginal cost function when maximizing profits using as strategic variable the produced quantity. The price function

$$p(x) = A - Bx + Cx^2 - Dx^3$$

Journal of Physics: Conference Series **692** (2016) 012013 doi:10.1088/1742-6596/692/1/012013

is kept monotonically decreasing by requiring that A, B, C and D are strictly positive and $C^2 < 3BD$. The revenue of the monopolist is

$$R(x) = p(x)x.$$

The marginal revenue is then given by

$$MR = \frac{\mathrm{d}R}{\mathrm{d}x} = p + x\frac{\mathrm{d}p}{\mathrm{d}x}.$$

The marginal cost is assumed to be

$$MC = E - 2Fx + 3Gx^2,$$

where the parameters E, F and G are positive constants. The profit is maximized when $MR = MC$ using standard results of economic theory. The profit function is

$$\Pi(x) = (A - E)x - (B - F)x^2 + (C - G)x^3 - Dx^4. \tag{2}$$

The profit function (2) is assumed to be not explicitly known by the monopolist. In [7] a simple Newton-Raphson-like algorithm is used to find the maximum of (2). The last two visited points x_t and y_t are evaluated and using a step size δ we obtain the next point y_{t+1} as

$$\begin{aligned} y_{t+1} &= y_t + \delta\tfrac{\Pi(y_t)-\Pi(x_t)}{y_t-x_t} \\ &= y_t + \delta\Big((A-E) - (B-F)(x_t+y_t) + \\ &\qquad (C-G)(x_t^2 + x_t y_t + y_t^2) - D(x_t^3 + x_t^2 y_t + x_t y_t^2 + y_t^3)\Big) \end{aligned} \tag{3}$$

The iteration of this procedure may lead to convergence to a profit maximum, an oscillating process or chaos depending on the value of the step size δ and the parameters. Following [7] we assume $A = 5.6$, $B = 2.7$, $C = 0.62$, $D = 0.05$, $E = 2$, $F = 0.3$ and $G = 0.02$. Using these parameter values, the profit function (2) is symmetric about the point $(3, 3)$. We keep the step size δ as a parameter.

The two step update process (3) can be interpreted as the two dimensional map:

$$M : \begin{pmatrix} x_t \\ y_t \end{pmatrix} \mapsto \begin{pmatrix} x_{t+1} \\ y_{t+1} \end{pmatrix} = \begin{pmatrix} y_t \\ y_t + \delta P(x_t, y_t) \end{pmatrix}, \tag{4}$$

with one parameter δ, where

$$P(x, y) = 3.6 - 2.4(x + y) + 0.6(x^2 + xy + y^2) - 0.05(x^3 + x^2y + xy^2 + y^3). \tag{5}$$

We refer to [2] for a detailed analysis of (4) which discusses the chaotic behavior of this monopoly model and investigates solutions of period 4, 5, 10, 13 and 17.

We implement (4) in MatContM for numerical analysis. To start a fixed point continuation we first need an initial fixed point.

The fixed points of (4) are obtained by solving

$$x = y, \tag{6}$$
$$y = y + \delta P(x, y). \tag{7}$$

to get

$$P(x, x) = 3.6 - 4.8x + 1.8x^2 - 0.2x^3 = 0. \tag{8}$$

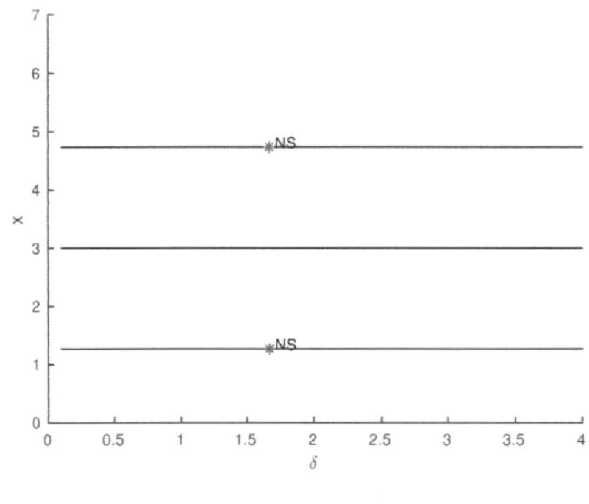

Figure 3: Results of fixed point continuations using the three fixed points of (4).

The fixed points of M are $x = 3 \pm \sqrt{3}$ (maxima of (2)) and $x = 3$ (minimum of (2)).

Next, we investigate the stability of these fixed points numerically using MatContM. We select a fixed point continuation and enter $x = y = 3 + \sqrt{3}$ in the `Starter` window and we select δ as the free system parameter with initial value 0.1. All other options remain on default. We select **Compute Forward** to start the continuation. We also enable the Numeric window for monitoring the multipliers. The results for all three fixed points are shown in Figure 3.

The fixed points $x = y = 3 \pm \sqrt{3}$ are initially stable for $\delta = 0.1$. The `Numeric` window indicates that the multipliers are within the unit circle. The fixed points lose stability around $\delta = 1.6666\ldots$ through a Neimark-Sacker (NS) bifurcation. This fixed point $x = y = 3$ is initially unstable and remains unstable throughout the continuation. This confirms the theoretical results in §2 of [2].

One of the remarkable properties of (4) is that a stable cycle of period-4 is born. We can obtain this cycle using simulation. We set $x = y = 4$ and $\delta = 1.8$. We select `Point` as an initial point type and select `Orbit (simulation)`. After computing hundreds of points we see the convergence to a period-4 cycle, see Figure 4. When we start from $x = y = 2$ we get another period-4 cycle.

We compute over 4000 points and select the last computed point. We can now continue the period-4 cycle by continuation of a fixed point of the iterated map $M^{(4)}$. Near $\delta \approx 2.62$ we start to detect many branch points (BP), this is caused by coexistence with other cycles with higher periods. In [2] it is also shown that the period-4 cycle loses stability at $\delta \approx 3.8647$.

Figure 5 repeats the computation of orbits for different values of δ. Notice how the fixed points evolve into period-4 cycles when δ passes the Neimark-Sacker bifurcation (upper-right). The two lower images show chaos setting in and we observe the two stable objects merge into a strange attractor.

MatContM allows to compute Lyapunov exponents. Select `Point` as `Point Type` and then select `compute Lyapunov exponents` as an initializer. The starter window allows to enter a range of parameter values. Figure 6 shows the largest Lyapunov exponents for δ between 1.5 and 4. A bifurcation diagram, obtained through simulation, is also included.

The negative values for σ in Figure 6 indicate the presence of stable cycles. Through simulations, we indeed find cycles of period 5, 10, 13 and 17 of (4). See Table 1. We then perform a numerical stability analysis. Once a cycle has been discovered for certain parameter value, it can be used as a starting point for continuation.

Journal of Physics: Conference Series **692** (2016) 012013 doi:10.1088/1742-6596/692/1/012013

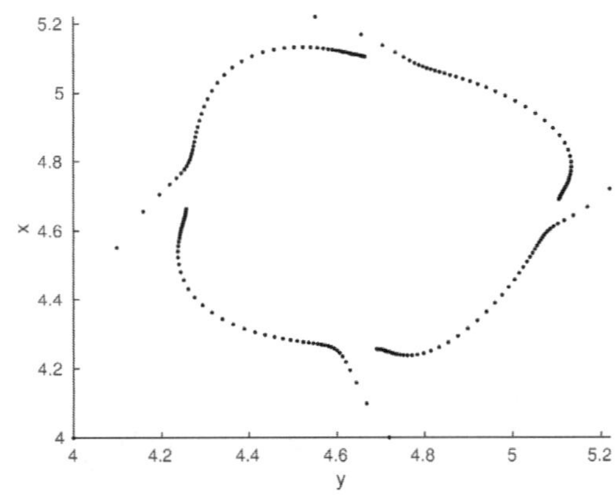

Figure 4: An orbit starting from $(4, 4)$ converges to a period-4 cycle when $\delta = 1.8$. We obtain ((5.10, 4.66), (4.66, 4.25), (4.25, 4.68), (4.68, 5.10)).

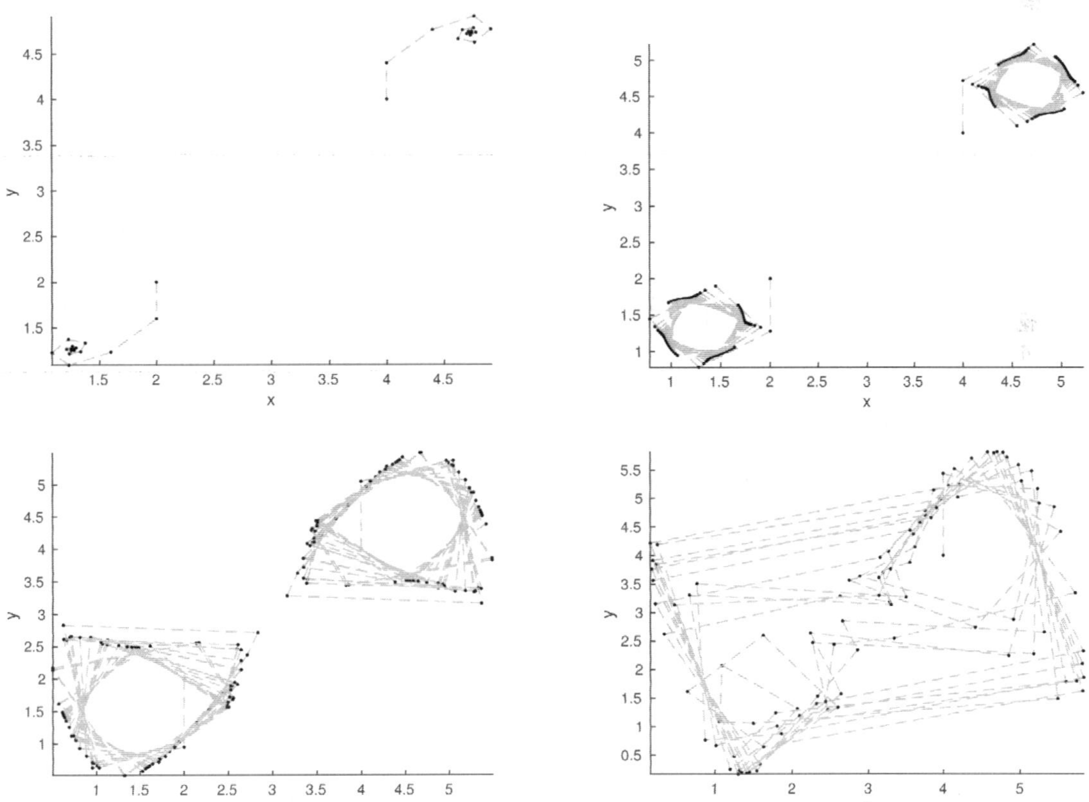

Figure 5: Orbits for $\delta = 1.0$ (upper-left), $\delta = 1.8$ (upper-right), $\delta = 2.62$ (lower-left) and $\delta = 3.6$ (lower-right). When $\delta = 1.0$ there is convergence to the stable fixed points. When $\delta = 1.8$ there is convergence to the stable period-4 cycles. For $\delta = 2.62$ and $\delta = 3.6$ we notice chaotic behavior due to coexistence with other cycles of different periods. The stable period-4 cycles persist but have a very small stability region.

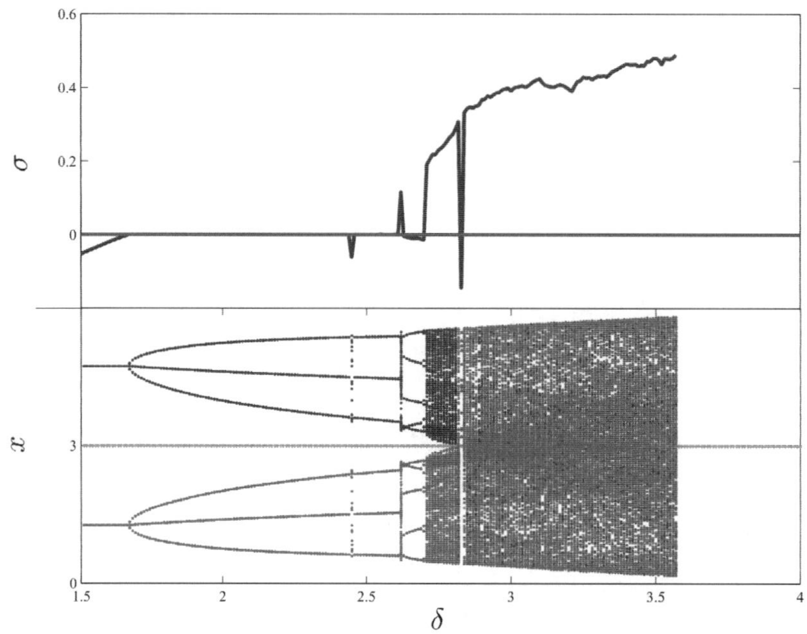

Figure 6: Top: the largest Lyapunov exponent, σ, as a function of δ. Negative values correspond to stable cycles, zero corresponds to 4-cycles with one multiplier equal to one, and positive values indicate chaotic behaviour. Bottom: bifurcation diagram obtained by simulation.

For the cycles of period 5, 10, 13 and 17, and using the initial data in Table 1, we continue each cycle with free parameter δ. The continuation of each cycle leads to a closed curve of cycles. Limit point (LP), Branch point (BP) and period-doubling (PD) bifurcations are found along these curves. Figure 7 shows the cycle of period 13 for $\delta = 2.83$, A part of the branch of 13-cycles is shown in Figure 8.

	5-Cycle	**10-Cycle**	**13-Cycle**	**17-Cycle**
δ	2.63	2.7	2.83	2.45
x	5.39438	3.99216	5.47699	4.24294
y	3.38445	4.86040	4.26739	5.16145

Table 1: One point on the 5,10,13,17-cycles.

During continuation, MatContM can also compute the multipliers which determine the stability of the 5, 10, 13 and 17-cycles.

Thus, the stability regions of the 5, 10, 13 and 17-cycles are bounded by the LP and PD points. The 10-cycles are stable in the regions bounded by the BP and PD points. The 5, 10, 13 and 17-cycles are unstable between two successive LP points or two successive PD points. Table 2 shows the stability regions for each cycle. The existence of PD bifurcations on the 10, 13, and 17-cycles indicates the existence of cycles of higher periods as well.

In *animation6.mpg* we show the continuation process of a 5-cycle in the state space and in

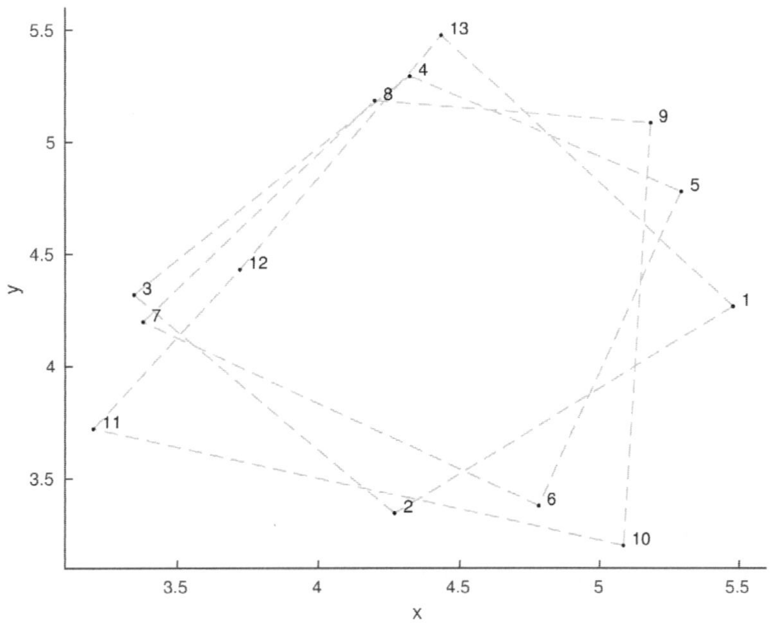

Figure 7: The period-13 cycle represented in Table 1 and used as initial data for computing Figure 8.

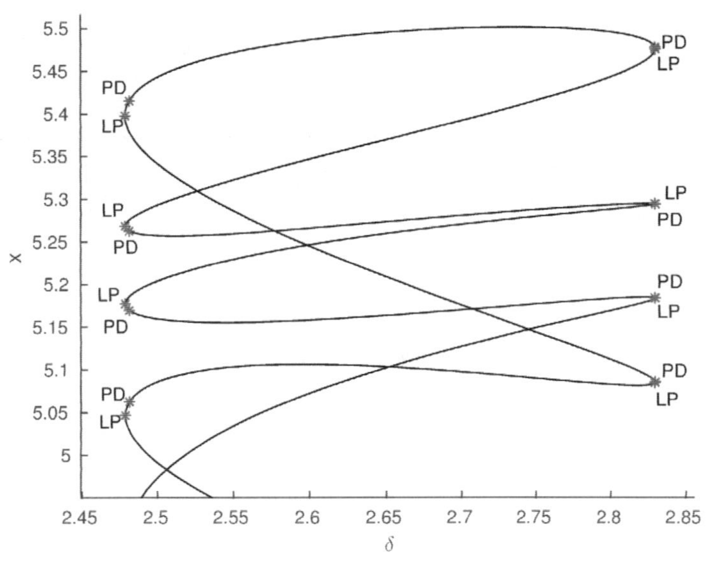

Figure 8: Zoom of a bifurcation diagram of period-13 cycles in (δ, x)-plane

the parameter space. We obtain a closed curve similar to Figure 8. We select a Period-Doubling bifurcation point and start a continuation of the doubled cycle of period 10. This continuation curve also has Period-Doubling bifurcations which allows us to start the continuation of a 20-

	Stable for δ in
5-Cycle	$(2.62813, 2.69111) \cup (3.52522, 3.52573)$
10-Cycle	$(2.69817, 2.70485) \cup (2.80134, 2.80214) \cup (3.12750, 3.12753) \cup (3.52149, 3.52522)$
13-Cycle	$(2.47864, 2.48136) \cup (2.82987, 2.83005)$
17-Cycle	$(2.44977, 2.45042) \cup (2.76425, 2.76432)$

Table 2: Stability regions of the 5,10,13 and 17-cycles of the monopoly model.

cycle.

Appendix A. Animations

The following video files were submitted:

(i) *animation1.mpg*: Continuation of a fixed point
(ii) *animation2.mpg*: Detection of bifurcations
(iii) *animation3.mpg*: Generating a bifurcation diagram
(iv) *animation4.mpg*: Continuation of connecting orbits
(v) *animation5.mpg*: Continuation of tangencies
(vi) *animation6.mpg*: Period-doubling of cycles

These files are also available on `http://sourceforge.net/projects/matcont/files/Documentation/MatcontM/files/NOMA15/`.

References

[1] H.N. Agiza, E.M. ELabbasy, H. EL-Metwally, and A.A. Elsadany. Chaotic dynamics of a discrete prey-predator model with Holling type II. *Nonlinear Analysis: Real World Applications*, 10(1):116 – 129, 2009.
[2] B. Al-Hdaibat, W. Govaerts, and N. Neirynck. On periodic and chaotic behavior in a two-dimensional monopoly model. *Chaos, Solitons & Fractals*, 70:27–37, 2015.
[3] A. Dhooge, W. Govaerts, and Yu. A. Kuznetsov. MATCONT: A matlab package for numerical bifurcation analysis of ODEs. *ACM Trans. Math. Softw.*, 29(2):141–164, June 2003.
[4] W. Govaerts, R. Khoshsiar Ghaziani, Yu. A. Kuznetsov, and H. G. E. Meijer. Numerical methods for two-parameter local bifurcation analysis of maps. *SIAM J. Sci. Comput.*, 29(6):2644–2667, October 2007.
[5] W. Govaerts, Yu. A. Kuznetsov, R. Khoshsiar Ghaziani, and H. G. E. Meijer. *Cl_MatContM: A toolbox for continuation and bifurcation of cycles of maps*, 2008. `http://sourceforge.net/projects/matcont`.
[6] R. Khoshsiar Ghaziani. *Bifurcations of Maps: Numerical Algorithms and Applications.* PhD thesis, Ghent University, 2008.
[7] T. Puu. The chaotic monopolist. *Chaos, Solitons & Fractals*, 5(1):35–44, January 1995.